T0296641

DIGITAL INNOVATIONS IN HEALTHCARE EDUCATION AND TRAINING

DIGITAL INNOVATIONS IN HEALTHCARE EDUCATION AND TRAINING

Edited by

STATHIS TH. KONSTANTINIDIS
University of Nottingham, Nottingham, United Kingdom

PANAGIOTIS D. BAMIDIS
Aristotle University of Thessaloniki, Greece

NABIL ZARY
Karolinska Institutet, Solna, Sweden

ACADEMIC PRESS

An imprint of Elsevier

Academic Press is an imprint of Elsevier
125 London Wall, London EC2Y 5AS, United Kingdom
525 B Street, Suite 1650, San Diego, CA 92101, United States
50 Hampshire Street, 5th Floor, Cambridge, MA 02139, United States
The Boulevard, Langford Lane, Kidlington, Oxford OX5 1GB, United Kingdom

Library of Congress Cataloging-in-Publication Data
A catalog record for this book is available from the Library of Congress

British Library Cataloguing-in-Publication Data
A catalogue record for this book is available from the British Library

ISBN: 978-0-12-813144-2

For information on all Academic Press publications
visit our website at https://www.elsevier.com/books-and-journals

Publisher: Stacy Masucci
Acquisitions Editor: Rafael Teixeira
Editorial Project Manager: Tracy Tufaga
Production Project Manager: Kiruthika Govindaraju
Designer: Miles Hitchen

Typeset by Thomson Digital

Working together
to grow libraries in
developing countries

www.elsevier.com • www.bookaid.org

Contents

Contributors

Panagiotis E. Antoniou
Medical Physics Laboratory, Medical School, Aristotle University of Thessaloniki, Thessaloniki, Greece

Panagiotis D. Bamidis
Aristotle University of Thessaloniki, Greece

Michael Campion
University of Washington, Seattle, WA, United States

Janet Corral
University of Arizona Tucson College of Medicine, Tucson, AZ, United States

Maria Foka
Department of Intensive Care Medicine, Nicosia General Hospital, Nicosia, Cyprus

Matěj Karolyi
Institute of Biostatistics and Analyses, Faculty of Medicine, Masaryk University, Brno, Czech Republic

Boyd Knosp
University of Iowa, Roy J. and Lucille A. Carver College of Medicine, Iowa, IA, United States

Martin Komenda
Institute of Biostatistics and Analyses, Faculty of Medicine, Masaryk University, Brno, Czech Republic

Stathis Th. Konstantinidis
University of Nottingham, Nottingham, United Kingdom

Theodoros Kyprianou
Department of Intensive Care Medicine, Nicosia General Hospital, Nicosia, Cyprus

Helen Macfarlane
University of Colorado School of Medicine, Aurora, CO, United States

Johmarx Patton
Association of American Medical Colleges, Washington, DC, United States

Eirini C. Schiza
Department of Intelligent Systems Group Johann Bernoulli Institute for Mathematics and Computer Science University of Groningen, Groningen, The Netherlands

Christos N. Schizas
Department of Computer Science, University of Cyprus, Nicosia, Cyprus

Pascal Staccini
Risk Engineering and Health Informatics Department, School of Medicine, Cote d'Azur University, Nice, France

Nicolas Stylianides
Department of Intensive Care Medicine, Nicosia General Hospital, Nicosia, Cyprus

Michael Taylor
School of Health Sciences, University of Nottingham, Nottingham, United Kingdom

Christos Vaitsis
Karolinska Institutet, Stockholm, Sweden

Heather Wharrad
School of Health Sciences, University of Nottingham, Nottingham, United Kingdom

Richard Windle
School of Health Sciences, University of Nottingham, Nottingham, United Kingdom

Neil Withnell
University of Salford, Manchester, England

Luke Woodham
St Georges's University of London, London, Great Britain

Nabil Zary
Karolinska Institutet, Solna, Sweden

Preface

Acquiring knowledge is a continuous process for people and especially for healthcare workforce. It is considered critical for doctors, nurses, and all the allied health professionals to be continuously updated, expand their knowledge, and trained in contemporary health-related competences, in order to provide high quality and effective care. The expansion of digital technology over the past decades provides additional means for education and training and it forms a new scientific field that continuously expands by incorporating a variety of digital innovations.

As the education and training of healthcare professionals evolve, the experts creating or co-creating digital innovations in healthcare education and training should combine a variety of competencies bringing together elements from pedagogy, technology, and health. Thus, this book aims to discuss and debate the contemporary knowledge about the evolution of digital education and learning and the role played by web and associated new technologies and approaches or services, their integration and the new pathways they create for modern healthcare education and training.

Taking into consideration that a creator of digital innovations in healthcare education and training should combine different multidisciplinary competences, this book has been written for people with different knowledge backgrounds and professions, including, but not restricted to, healthcare tutors, clinicians, pedagogists, computer scientist, and engineers that aim to learn more about digital innovations in healthcare education and training and come together to form multidisciplinary teams to create such advancements. In fact, academics involved in healthcare education will find it useful for updating their course material. On the other hand, it can be used by experts in the field that wish to get insights on the latest innovations. Furthermore, it can be a valuable resource for policy makers and managers to understand and value the need of digital innovation in healthcare education and training.

This book would not have been completed, without the contributions of the individual book chapters' authors, who were eager to participate to this endeavor and share their knowledge and work. Despite the fact that editing this book was a long process with many revisions, we believe that the end result brings together our initial vision to create a unique book that encompasses introductory knowledge, planning and designing digital innovations, state-of-the-art implementations, applied digital innovations in practice and identification of impact and sustainability of such advancements.

With the hope that this book will be a useful asset in your hands, we kindly invite you to read it and be inspired to co-create new digital innovations for healthcare education and training.

The Editors,
Stathis Th. Konstantinidis
Panagiotis D. Bamidis
Nabil Zary

PART ONE

Digital Innovations - A primer

CHAPTER ONE

Introduction to digital innovation in healthcare education and training

Stathis Th. Konstantinidis[a], Panagiotis D. Bamidis[b], Nabil Zary[c]
[a]University of Nottingham, Nottingham, United Kingdom
[b]Aristotle University of Thessaloniki, Greece
[c]Karolinska Institutet, Solna, Sweden

Chapter outline

Introduction

Information and communication technology (ICT) has a rapid expansion within the last decades along with the continuous evolution of healthcare and medicine evidence and techniques. Furthermore, technology started many decades ago to be used in healthcare education and training not only to solve problems of practical training, like simulations but also enhance the knowledge acquisition holistically. Thus, the most advanced digital innovations are continuously embedded in healthcare education and training. *Digital innovation* includes either new technology or novel use of an existing technology; with technology refer to both software and hardware. *Healthcare education and training* encompasses notions of learning, teaching, evaluating, the process of acquiring knowledge, skills and competencies, and even behavior modification through different types of exercises, including active and passive learning methods.[1] The uniqueness of healthcare education and training requires the learners to be up-to-date with the latest advancements in their respected field, in an efficient way, around the tight schedules of both students and professionals.

Digital Innovations in Healthcare Education and Training.
http://dx.doi.org/10.1016/B978-0-12-813144-2.00001-5

Pedagogy and online pedagogy

A variety of learning and teaching theories have been widely used in healthcare education and training, in different settings, to enhance a range of competencies and skills. Healthcare education and training considered a very active area of educational research.[2]

One of the main things that have become central to healthcare education and training is the intrusion of social constructivism,[3] which is widely adopted nowadays. Social interaction with other tutors and learners is fundamental in the development of understanding. Learners create their knowledge based on what they already know, and they are choosing what is important to learn to enhance their competences. Thus, the tutor-centric model has been moved toward learner-center teaching and learning approach.[4] Lave and Wenger[5,6] proposed the term Community of Practice to show the importance of integrating individuals into the community and the role of the community on individual practices. Learners are creating their overlapping knowledge of practice and at the same time and focus on common experiences. Healthcare education uses the four steps of experiential learning[7] since the transition of a student to an active healthcare professional relies progressively more on experiential learning.[8]

Healthcare education and training is considered an early adopter of innovative methodologies and theories and digital technologies could not be left aside. Healthcare education and training envisages the use of multiple educational approaches to enhance learning, and the use of technology is done alongside these lines.

Co-creation value

Over the last decades, there was a tendency to move closer to the future users of digital applications.[9,10] Similarly, user-centered approaches for developing educational resources become more common. The practice of collective creativity in design, known as participatory design is around for many decades.[11] Nowadays, there is a shift from user-centered design to co-design with the stakeholders that will use the final educational resources,[12] and even in some cases co-create the final educational resources, as they do not only participate in the design of the digital education resource but in the production of its different assets. Following these changes, the researcher or tutor alter her role form translator between the "learners" and designer, to facilitator for the stakeholders (learners) that co-design or co-create the re-source.[9] Simultaneously, many co-creation and co-design methodologies have been proposed.[12-16] Co-creation of digital resources value proved with evidence both with perceptions of users of the digital resources on them,[17,18] but also with evidence showcasing the contribution of these digital resources on the improved knowledge of the users following these resources.[19]

According to Datta et al.[20] simulation is *"the artificial representation of a complex real-world process with sufficient fidelity with the aim to facilitate learning through immersion, reflection, feedback, and practice minus the risks inherent in a similar real-life experience."* Simulation-based education has its place in health-care education, and at the infancy of digital healthcare education, simulation was considered a synonym. Virtual Patients[21-23] is a form of interactive computer simulations of real-life clinical scenarios for medical training, education, or assessment. Virtual Patients enable learners to explore different decision paths and learn through the proses of one-to-one consultation with the virtual patient. Similarly, serious games simulations provide a more real environment for the learner. Debriefing such scenarios allow the learner to discuss further and receive personalized feedback that leads to the acquisition of in-depth knowledge on the topic areas that either she or the facilitator feels is of high importance.[24-27]

 ## Innovations coined the digital innovation in healthcare education and training (research) area

Collaborative Web or Web 2.0 brought together in an online community concept of active participation and allowed virtual places or spaces for learners and experts to collaborate, explore and create new knowledge and expertise, enhancing learning experience.[28] Many different collaborative tools have been used for healthcare education and training, such as Wikis, Blogs, Mashups, Virtual Words, Microblogs, and others,[29] which can lead the learners to:[30]

- effortless communication and collaboration among peers;
- ample access to alternative sources of information, usually customarily combined at a meta-level;
- reporting and rating of information within open communities of peers;
- potential for different representations of the same content (for people with special needs, with different cultural backgrounds, different ages, different background); and
- context-based organization of resources and activities.

Examples of the later one included the sharing of educational resources through social networks, but also linking the educational resources as a big resources network based on location, relationship of owners, themes,[31] re-purposing, and inheritance.[32]

Semantic Web or Web 3.0 rose the last decade, even if it has been proposed years before,[33] and in parallel digital innovation encompassing the use of it have been proposed ranging from sharing and retrieving educational resources based on their semantic descriptions,[34] to automatically semantically describe and tailor to the learner needs existing resources, such as YouTube Videos[35] and Wikis,[36] or more advanced Virtual Patients.[23]

Sharing of resources and systematic recording of health education related education could not be achieved if there were a list of education specifications and standards to systematic record description around the educational process.[37–41]

Such standards allow the notion of vast silos of related educational information including activity data. Moreover, even if such information is not always possible, decision making systems and learning analytics allows not only healthcare learners and educators to benefit from their analysis, but also might provide valuable results for educational organizations and policymakers.[42,43]

Healthcare education and training should be also interface-friendly and tailored to users' needs. Usability is core to such developments,[44] while persuasive technologies should be always considered.[45] Different type of usability monitoring such as eye-tracking[46,47] or monitoring of website are used.[48–50].

All the previous-mentioned notions allows for the term **Medical Education Informatics** to be coined by Bamidis[51] and embraces all the digital advances in the field of Digital Innovations on Healthcare Education and Training, bringing together researchers and educators in the multidisciplinary areas of medical/health education; pedagogy; openness; education technology; Social (WEB2.0), and Semantic Web (Web3.0); technology enhanced learning (Fig. 1.1).

Book structure

This book aims to discuss and debate the contemporary knowledge about the evolution of digital education and learning and the web and its integration and role within modern healthcare education and training. The book encompasses topics such as healthcare and medical education theories

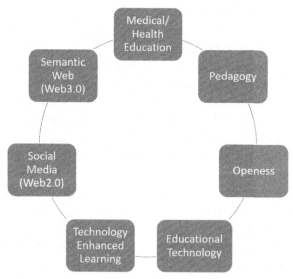

Figure 1.1 *Digital innovations on healthcare education and training area.*

and methodologies, social learning as a formal and informal digital innovation, and the role of semantics in digital education. The book also examines how simulation, serious games, and virtual patients change the way of learning in healthcare, and how learning analytics and big data in healthcare education leads to personalized learning. Online pedagogy principles and applications, participatory educational design, and educational technology as health intervention are bridged together to complement this collaborative effort.

This book is a valuable resource for a broad target of audience both technical and non-technical including healthcare and medical tutors, health professionals, clinicians, psychologists, web scientists, engineers, computer scientists, managers, academics, and any other relevant professional, interested in using and creating digital innovations for healthcare education and training.

The book aims to meet the following needs and challenges per target audience:

- Healthcare and medical tutors: What are the current digital innovations that I can use in my teaching and content development? How can I use them? Where can I find such examples?
- Health professionals, clinicians, psychologists (beyond the role of a tutor): How digital innovations in healthcare education and training influence

my healthcare practice? What are the ways to be kept up-to-date with the latest advancements in my field?

- Web scientists, learning technologists, engineers, computer scientists? What are the latest technologies in this topic? What is the big picture behind the technology? What is the role of data in healthcare education and training? What are the best practices and principles to use for developing applications and systems for healthcare education?
- Managers, policymakers: What technologies are available and how they can increase the engagement and satisfaction of the learners and tutors? What is the IT literacy needed for the healthcare workforce in order to be trained online? What are the digital innovation in healthcare education and training for managers and policymakers?
- Academics, researchers: What are .the current trends in the topic? How can I use digital innovation to increase personalized learning and transfer the latest research results into education and training? How can I evaluate digital innovations in healthcare education or use them as health interventions?

Thus, the remainder of the book has the following structure: Initially, two introductory chapters introduce us to the basics of digital innovation in healthcare education together with the first one.

Chapter 2: *Serious games, simulations, and virtual patients* enable us to understand computer-based applications are increasingly used to support the training of health care professionals for initial and continuous teaching. By means of virtual patients scenarios embedded in augmented or virtual reality contexts, played through serious games or simulation sessions, students and teachers face new educational opportunities without clear evidence of effectiveness on the overall pedagogical process (teaching and learning). This chapter presents the main digital innovations (augmented reality, virtual reality, virtual patients, and serious games) according to the most recent reviews published in medical literature. The question of knowledge retention and student performance remains the core background to be explored. The concept of "digitally augmented curricula" seems to be the starting point for defining best practices and review the roles, the utility, the usage, and the risks/errors of digital tools for medical education.

Chapter 3: *Designing digital education and training for health* introduce us to co-creation approaches. In this chapter some of the underpinning learning theory and instructional design models that support the design of digital education and training for health will be explored. The theories, and development and design methodologies could equally apply to digital learning resources, modules or whole courses. However, the focus here will be on a digital learning resource format know as a reusable learning object (RLO) which as the name suggests is characterized as a short (10 min of learning), focused, self-contained digital resource covering a discrete topic. An example of a development methodology incorporating co-production, peer review and pedagogical design, for digital learning resources for healthcare student education, patient education and in the training of healthcare professionals (HCPs) will be described and demonstrated using three case studies.

The next section aims to highlight the digital innovation into education and training, but also considerations while educators are planning and designing digital innovation to enhance healthcare education and training.

Chapter 4: *The role of educational (technical) standards in healthcare education and training* introduces us to technical standards in medical and healthcare education, which are usually developed and applied to systematically enhance continuous improvements of teaching/learning processes. In general, the actual use of these standards in practice ensures both interoperability and reusability of various educational objects and tools together. Moreover, standardization helps to avoid custom-made solutions and proprietary software products that are supposed to improve the quality of tertiary education. Technical recommendations and guidelines correctly link up the medical informatics domain with good healthcare practices under the supervision of professional organizations and communities from around the world. This chapter introduces the usage of selected technical standards in medical education from three of today's most up-to-date and discussed perspectives: curriculum development and mapping, virtual patients and user behavior, and assessment of student learning.

Chapter 5: *Internet of Things in healthcare education and training* explains how the real environment can enhance healthcare education. Healthcare education is continuously looking for new ways of delivering learning and teaching. Recently the Internet of Things (IoT) came into the foreground envisioning a future in which digital and physical entities can be linked. While the idea of pervasive, context-aware and location-based game learning is not new, they all can be included under the IoT in education notion, as a digital innovation enhancing healthcare education. This chapter debates the role of IoT in education in a healthcare context. Initially, a discussion and a definition of what considered as IoT will be given, followed by an overview of IoT in healthcare. Next, the technologies used in IoT are briefly discussed, followed by an analysis of security issues and challenges for IoT.The theory underpins the IoT in education is explained, succeed by a discussion on the benefits and limitation of a context-aware educational system interacting with the environment or location-based learning games. The case ofViRLUS is presented as a representative combination of current and future digital innovations. Finally, yet importantly, the IoT in education concepts are summarized, and future visions are discussed.

Chapter 6: *Social web and social media's role in healthcare education and training* introduces us to social media and the role that they can play in healthcare education providing also the challenges that they meet in real healthcare environments.The use of social media is growing apace, and there is undoubtedly potential in using social media in healthcare, and in healthcare education. An organization can quickly "spread the word" and use social media to both inform and protect, the public. Caution must be taken to ensure that information is not misleading and professionals must abide by their regulatory body.This chapter will explore the use of social media in healthcare.

The next section is implementing digital educational innovation for health, in which state-of-the-art implemented concepts are thoroughly discussed.

Chapter 7: *Implementing digital learning for health* explores the implementation of different existing good paradigms using pivotal technologies. The single most significant challenge in contemporary medical education is the absorption, by the students of a rapidly expanding volume of medical knowledge in ways supporting critical thinking and initiative.There is a severe educational disparity between theoretical knowledge and clinical skills or decision–making capacity in medical education. Immersive and interactive, experiential digital education episodes provide impactful learning that conveys both formal as well as tacit knowledge through meaningful narratives.This chapter presents a brief rationale and implementation modalities for such experiential narratives. It explores endeavors for integrating virtual patients in various experiential modalities such as the 3D multi-user virtual environments (MUVEs), as well as virtual and augmented reality ones. Exploring future directions, this chapter closes with points about the necessary data modeling provisions, as well as the conceptual paradigm shift integrating these educational endeavors into a versatile, co-creative educational living lab environment for immersive experiential and impactful medical learning.

Chapter 8: *Artificially intelligent Chatbots for health professions education* enables the reader to understand the role that Chatbots can play in healthcare education and training. Chatbots are computer software designed to have conversations with humans. Chatbots are widely used in the customer service industry, where frequently asked questions or basic information is answered by the computer. Chatbots may serve useful functions for supporting health professions trainees across geographically distributed sites, as well as reducing faculty burnout. This chapter explores the potential roles of Chatbots to help with scaling faculty and student time within fast-moving health professions programs. The possibilities for Chatbots are organized with Bloom's taxonomy to gain an appreciation for the simple through advanced applications. Building upon these ideas, the chapter also introduces the VoxScholar project to address how artificially intelligent Chatbots might coach to teaching expertise or academic success.

Chapter 9: Learning analytics, data mining, and personalization discusses both the technical and the theoretical concept to enable personalization by understanding activity data. Personalization of student evaluation data and learning application usage has the potential to provide targeted feedback to support self-directed learning and expertise development among learners in health professions education. To provide personalization, health professions programs need to leverage both: (1) the technical infrastructure and analysis for existing student data, and (2) understand the interrelated contexts of learning, teaching, and expertise development within the clinical setting. Personalization is more than feeding back results to learners; it also moves into intelligent tutors. Tips for successful adoption of personalization in health professions education, including security, legal, and administrative concerns, are discussed.

The last part of this book provides examples of how these technologies have been applied in practice, how they can be evaluated and become sustainable digital innovation.

Chapter 10: *What's in your medical education data warehouse? Results from interviews with 11 US medical schools.* Education data warehouses (EDWs) are an important and foundational information technology infrastructure of a successful program of learning analytics. EDWs, by virtue of their name, acknowledge that education data is not just within one database, but usually one location—the warehouse—that connects multiple data sets distributed across digital systems. Within medical schools, while the concept of an EDW may be grasped with an increasing level of certainty, the complexity of creating and implementing EDWs and using data from EDWs, represents a significant task involving many steps that are, at present, poorly defined. Accordingly, the authors interviewed a range of new and established EDW adopters from 12 medical schools across the United States to better understand the contexts of use and emerging best practices. This chapter presents the summary data and analysis resulting from those semi-structured interviews, toward informing practice across the medical school community.

Chapter 11: *Teaching and integrating ehealth technologies in undergraduate & postgraduate curricula and healthcare professionals' education & training* provides the opportunity to showcase some of the proposed tools into real curricula, either as existing or future digital innovations. This book chapter offers a new insight into the methodology of teaching eHealth by integrating eLearning tools at the undergraduate medical level, the postgraduate level, and at the level of continuing professional education (CPE) in highly demanding clinical environments such as that of critical care. Many and significant challenges are posed to a healthcare profession student today, at both undergraduate and postgraduate levels, as one is required to learn and practice or even design the modern and technology-rich clinical environment. Even more challenges are faced by the healthcare professional that is responsible for "real patients" and require taking decisions on the job and occasionally under high pressure. Such decisions must be based on accurate and reliable data, complete, and readily available. Conclusively, we discuss prospects of learning and practicing eHealth, the challenges in integrating innovative IT technologies to eLearning and the concept of embedding those processes to a real-time assess-educate-assess cycle that uses real-time data analytics and advanced micro-learning tools to optimize outcome.

We believe that this book will provide contemporary knowledge about the evolution of learning technologies and the web and its integration and role within modern healthcare education and training. It will inform its reader with the latest digital innovation in healthcare education and training enabling all type of readers to apply them in their practice, being tutor, learner, developer, manager, or policy maker. As a useful paradigm of a cross-theme scholarly explanation on digital innovation in healthcare education and training, it provides an in-depth knowledge experience into digital innovation in healthcare education and training.

References

1. Masic I, Pandza H, Toromanovic S, et al. Information technologies (ITs) in medical education. *Acta Inform Med* 2011;**19**:161–7.
2. Flores-Mateo G, Argimon JM. Evidence based practice in postgraduate healthcare education: a systematic review. *BMC Health Serv Res* 2007;**7**:119.
3. Vygotsky L. *Mind in society: the development of higher psychological processes.* Cambridge: Harvard University Press; 1978.

4. Bamidis PD, Konstantinidis ST, Kaldoudi E, et al. New approaches in teaching medical informatics to medical students. In: *Twenty-first, IEEE, international symposium on computer-based medical systems, IEEE*; 2008. p. 385–390.

5. Lave J, Wenger E. *Situated learning: legitimate peripheral participation.* Cambridge, UK: Cambridge University Press; 1991. DOI: 10.2307/2804509.

6. Wenger E. Communities of practice and social learning systems: the career of a concept. In: *Social learning systems and communities of practice*; 2010. DOI: 10.1007/978-1-84996-133-2_11.

7. Kolb DA. *Experiential learning: experience as the source of learning and development*; 1984. DOI: 10.1016/B978-0-7506-7223-8.50017-4.

8. Yardley S, Teunissen PW, Dornan T. Experiential learning: transforming theory into practice. *Med Teach* 2012;**34**:161–4.

9. Sanders EB-N, Stappers PJ. Co-creation and the new landscapes of design. *CoDesign* 2008;**4**:5–18.

10. Bowen S, Dearden A, Wright P, et al. Participatory healthcare service design and innovation. In: *Proceedings of the eleventh Biennial Participatory Design Conference on - PDC '10*; 2010. DOI: 10.1145 / 1900441.1900464.

11. Cross N, editor. *Design participation: proceedings of the Design Research Society's conference 1971.* London, UK: Academy Editions, 1972, https://catalogue.nla.gov.au/Record/1916512.

12. Mor Y, Winters N. Participatory design in open education: a workshop model for developing a pattern language. *J Interact Media Educ* 2008;**2008**(1):12 DOI: 10.5334/2008-13.

13. Winters N, Mor Y. IDR: a participatory methodology for interdisciplinary design in technology enhanced learning. *Comput Educ.* 2008;**50**:579–600 DOI: 10.1016/j.compedu.2007.09.015.

14. Kensing F, Simonsen J, Bodker K, et al. MUST: a method for participatory design. *Hum Comput Interact* 1998;**13**:167–98.

15. Boyle T, Cook J, Windle R, et al. Agile methods for developing learning objects. In: *Proceedings of the 23rd Ascilite Conference.* Ireland, https://www.semanticscholar.org/paper/An-Agile-method-for-developing-learning-objects-Boyle-Cook/3830247d23e3f653039271f4c8a6b199575ca3d6.

16. Windle R, Wharrad H, K C, et al. Collaborate to create; Stakholder participation in Open content Creation. In: *Presented at Association for Learning Technology Conference (ALT-C).* Warick, 2016.

17. Lymn JS, Bath-Hextall F, Wharrad HJ. Pharmacology education for nurse prescribing students—a lesson in reusable learning objects. *BMC Nurs* 2008;**7**:2.

18. Blake H. Computer-based learning objects in healthcare: the student experience. *Int J Nurs Educ Scholarsh*; 7. DOI: 10.2202/1548-923X. 1939.

19. Windle RJ, McCormick D, Dandrea J, et al. The characteristics of reusable learning objects that enhance learning: a case-study in health-science education. *Br J Educ Technol* 2011;**42**:811–23.

20. Datta R, Upadhyay KK, Jaideep CN. Simulation and its role in medical education. *Med J Armed Forces India*; 2012. DOI: 10.1016/S0377-1237(12)60040-9.

21. Poulton T, Balasubramaniam C. Virtual patients: a year of change. *Med Teach* 2011;**33**:933–7.

22. Zary N, Johnson G, Boberg J, et al. Development, implementation and pilot evaluation of a Web-based Virtual Patient Case Simulation environment--Web-SP. *BMC Med Educ* 2006;**6**:10.

23. Dafli E, Antoniou P, Ioannidis L, et al. Virtual patients on the semantic Web: a proof-of-application study. *J Med Internet Res* 2015;**17**:e16.

24. Georg C, Henriksson EW, Jirwe M, et al. Debriefing of virtual patient encounters: systematic collection of nursing students clinical reasoning activities. DOI: 10.7287/peerj. preprints.1172v1.

25. Arafeh JMR, Hansen SS, Nichols A. Debriefing in simulated-based learning: facilitating a reflective discussion. *J Perinat Neonatal Nurs* 2010;**24**:302–9.
26. Dreifuerst KT. The essentials of debriefing in simulation learning: a concept analysis. *Nurs Educ Perspect* 2009;**30**:109–14.
27. Paige JT, Arora S, Fernandez G, et al. Debriefing 101: training faculty to promote learning in simulation-based training. *Am J Surg* 2015;**209**:126–31.
28. Boulos MNK, Maramba I, Wheeler S. Wikis, blogs and podcasts: a new generation of Web-based tools for virtual collaborative clinical practice and education. *BMC Med Educ* 2006;**6**:41.
29. Kaldoudi E, Konstantinidis S, Bamidis PD. *Web advances in education: interactive, collaborative learning via web 2.0;* 2010. DOI: 10.4018/978-1-60566-940-3.ch002.
30. Kaldoudi E, Konstantinidis S, Bamidis PD. *Web 2.0 approaches for active, collaborative learning in medicine and health;* 2010. DOI: 10.4018/978-1-61520-777-0.ch007.
31. Konstantinidis ST, Dovrolis N, Kaldoudi E, et al. Geotagged repurposed educational content through mEducator social network enhances biomedical engineering education. In: *IFMBE Proceedings;* 2010. DOI: 10.1007/978-3-642-13039-7_245.
32. Kaldoudi E, Dovrolis N, Konstantinidis ST, et al. Depicting educational content repurposing context and inheritance. *IEEE Trans Inf Technol Biomed* 2011;**15**:164–70.
33. Berners-Lee T, Hendler J, Lassila O. The semantic Web. *Sci Am* 2001;**284**:34–43.
34. Konstantinidis ST, Ioannidis L, Spachos D, et al. mEducator 3.0: combining semantic and social web approaches in sharing and retrieving medical education resources. In: *In Proc of 2012 seventh international workshop on semantic and social media adaptation and personalization (SMAP 2012).* Luxemburg, 2012, pp. 42–47.
35. Konstantinidis S, Fernandez-Luque L, Bamidis P, et al. The role of taxonomies in social media and the semantic Web for health education. a study of SNOMED CT terms in YouTube health video tags. *Methods Inf Med* 2013;**52**:168–79.
36. Bratsas C, Kapsas G, Konstantinidis S, et al. A semantic wiki within moodle for Greek medical education. In: *2009 22nd IEEE international symposium on computer-based medical systems.* IEEE, p. 1–6.
37. Konstantinidis ST, Kaldoudi E, Bamidis PD. Enabling content sharing in contemporary medical education: a review of technical standards. *J Inf Technol Healthc* 2009;**7**:363–75.
38. Vaitsis C, Spachos D, Karolyi M, et al. *Standardization in medical education: review, collection and selection of standards to address technical and educational aspects in outcome-based medical education;* 2017. Facta Medica, http://mj.mefanet.cz/mj-20170523.
39. Hersh WR, Bhupatiraju RT, Greene P, et al. Adopting e-learning standards in health care: competency-based learning in the medical informatics domain. *AMIA. Annu Symp proceedings AMIA Symp* 2006;**2006**:334–8.
40. Bamidis PD, Nikolaidou MM, Konstantinidis ST, et al. A proposed framework for accreditation of online continuing medical education. In: *Twentieth IEEE international symposium on computer-based medical systems (CBMS'07).* IEEE, p. 693–700.
41. Vaitsis C, Stathakarou N, Barman L, et al. Using competency-based digital open learning activities to facilitate and promote health professions education (OLAmeD): a proposal. *JMIR Res Protoc* 2016;**5**:e143.
42. Konstantinidis ST, Bamidis PD. Why decision support systems are important for medical education. *Heal Technol Lett* 2016;**3**:56–60.
43. Konstantinidis ST, Fecowycz A, Coolin K, et al. A proposed learner activity taxonomy and a framework for analysing learner engagement versus performance using big educational data. In: *2017 IEEE thirtieth international symposium on computer-based medical systems (CBMS);* 2017, p. 429–434.
44. Sandars J, Lafferty N. Twelve tips on usability testing to develop effective e-learning in medical education. *Med Teach* 2010;**32**:956–60.

45. Bamidis PD, Konstantinidis ST, Bratsas C, et al. Federating learning management systems for medical education: a persuasive technologies perspective. In: *Twenty-fourth international symposium on computer-based medical systems (CBMS)*. IEEE 2011, p. 1–6.
46. Lai ML, Tsai MJ, Yang FY, et al. A review of using eye-tracking technology in exploring learning from 2000 to 2012. *Educ Res Rev* 2013;10:90–115.
47. Was C, Sansosti F, Morris B. *Eye-tracking technology applications in educational research*; 2016.
48. Çetin E, Özdemir S. A study on an educational website's usability. *Procedia Soc Behav Sci* 2013;**83**:683–8.
49. Antoniades A, Nicolaidou I, Mylläri J, et al. Design of evaluation of content sharing solutions in medical education. In: *Fourth international conference of education, research and innovations*. Madrid, Spain, 2011, p. 6786–6794.
50. Vosylius AE, Lapin K. Usability of educational websites for tablet computers. In: *Proceedings of the mulitimedia, interaction, design and innovation on ZZZ—MIDI '15*. New York, New York, USA: ACM Press, p. 1–10.
51. Bamidis PD. Medical/health education informatics: from birth to grow; an account of achievements. In: Bamidis PD, Konstantinidis ST, (editors). *Medical education informatics 2018 (MEI2018)*. Thessaloniki, 2018.

CHAPTER TWO

Serious games, simulations, and virtual patients

Pascal Staccini

Risk Engineering and Health Informatics Department, School of Medicine, Cote d'Azur University, Nice, France

Chapter outline

Nowadays, virtual reality is part of our daily life as citizen or professional, in the form of representations, animation films, or simulators. Simulation is a technique currently being used by many educators in a variety of fields. Training tools have been developed for various professions, more or less at risk, in the military, aviation, nuclear power industries, or business and management, as part of their overall training and readiness programs. Henceforth, health care facilities, medical, midwifery, and nursing schools have incorporated simulation in curricula, in an effort to enhance learning related to procedural training, team training, and individual learner training. In health care, a key benefit for using simulation is its ability to mimic real life situations without putting patients at risk. The methods for representing artificial or augmented environments are continually improved. Research teams are developing models that are approaching more and more accurately the world of life, at several levels of scale, with the ability to estimate and integrate many parameters. Thus, realistic 3D environments combining refined ergonomics and scenarios based on real cases, offer a new reality: allowing professionals to gain command of a product even before its existence, to train hand interaction with virtual objects in real-time while mimicking tasks from real life, or to practice care management in quasi-real conditions (immersive experience), to acquire and successfully implement knowledge and clinical expertise, and thus reduce errors done after training. Simulation is a practical and successful model which can be used to teach a variety of skills: psychomotor (technical), cognitive (clinical reasoning, decision making), and interpersonal (communication, teamwork). In the world of medical interactive simulations, there are three classifications of simulations used: low–fidelity (non-computerized trainers that teach a specific task

Digital Innovations in Healthcare Education and Training.
http://dx.doi.org/10.1016/B978-0-12-813144-2.00002-7

such as intravenous catheter insertion), mid-fidelity (standardized patients, computer programs, video games), and high fidelity (computerized human patient simulator mannequins that respond to treatment). Simulation-based training can be ideally considered as a comprehensive training platform[1] founded on case-based learning (CBL), in the form of team-based learning (TBL), in the space of scenario-based learning (SBL), directed by problem-based learning (PBL), with the final goal of problem solving-based learning (PSBL). In a recent review of the design and development processes of simulation for training in healthcare, Persson[2] used the following terms to represent "simulation systems": simulation, visualization, "virtual reality," "virtual environment," mannequin, "desktop simulation," "computer-based scenarios," "simulated patients," "virtual patients," "online worlds," game. This means that speaking about digital innovations for simulation, we have explored four digital entities: augmented reality (AR), virtual reality (VR), virtual patient (VP), and serious game (SG).

Augmented reality differs from its most known "relative" virtual reality, since the latter creates a 3D world completely detaching the user from reality. There are two aspects in which augmented reality is unique: users do not lose touch with reality and it puts information into eyesight as fast as possible. Measuring the effectiveness of virtual and augmented reality in health sciences and medical anatomy, Moro et al.[3] defined (AR) as: "Using a camera and screen (i.e., smartphone or tablet) digital models are superimposed into the real-world. The user is then able to interact with both the real and virtual elements of their surrounding environment." Despite (augmented reality) is not an entry term in MeSH classification, we found 1077 current references (Nov. 2017). First ones were indexed in 1995 but focused on virtual guidance during surgical operation.[4] Zhu et al.[5] provided in 2014 an extensive review of AR in healthcare education and categorized papers according to subjects, aim, role of AR, targets, and computer systems. Refining the initial query with (training or learning), 289 references were eligible, first indexed in 1997[6] about visualization of dynamic 3D anatomy. The last reference to be indexed[7] describes an "augmented mirror" to learn anatomy. Because the cadaver-based learning approach has seen a decline due to practical, ethical, and cost issues, anatomical education has relied more on physical, diagram, and image models. The "augmented mirror" approach introduces additional spatial learning challenges for students, including difficulties in (1) translating two-dimensional static images, diagrams, or photographs in medical illustrations to 3D human bodies and (2) visualizing dynamic processes in a living organism, such as the activity of working muscles, from inanimate physical models. The system consists of an RGB-D sensor as a real-time tracking device (Microsoft Kinect), which enables the system to link a deposited section image to the projection of the user's body, as well as a large display mimicking a real-world physical mirror. Using gesture input, the users have the ability to interactively explore radiological images in different anatomical intersection planes (high resolution photographic images from the Korean Phantom Dataset). Skeletal information can be extracted, in particular the 3D positions of joints, from the active user by means of machine learning algorithms. The system displays anatomical slices from various modalities as well as different section planes to the user and interactively switches between them[8]. 800 first-year medical students tested and evaluated the system. Emphasis was put on active learning, 3D understanding, and a better comprehension of the course of structures. These results seem to confirm Moro et al.'s conclusion[3] that AR (as well as VR) provides intrinsic benefits, such as increased student engagement, interactivity, and enjoyment, but does not necessarily result in increased test scores. Additional benefits have been reported by Zhu et al.[5] such as: more knowledge acquired, skill acquisition improvement, learning retention, shorten learning curve, and better understanding of spatial interrelationships. Although Zhu et al. owned up their review was done in the early stages of AR, while most tested applications were prototypes, they noticed the lack of learning theories to guide the design of AR. In 2016, Barsom et al.[9] investigated to which extent augmented reality applications were used to validly support medical professionals training. They classified applications in three different categories: laparoscopic surgical training, mixed reality training of neurosurgical procedures, and training echocardiography. They concluded that, if several applications have shown the potential of AR to bridge the gap between achieving the actual competence needed in the real working environment and training them in a virtual context, the implementation of existent and new AR in a training curriculum of medical specialists, needs assessment strategies and complete validation trajectory.

In a virtual reality system,[3] the user's senses (sight, hearing, and motion) are fully immersed in a synthetic environment that mimics the properties of the real world through high resolution, high refresh rate head-mounted displays (HMD), stereo headphones, and motion-tracking systems. This enables users to become fully immersed in a simulated world. The use of the HMD helmet allows users to perceive 3D stereoscopic images and to determine the spatial position in the visual environment via motion tracking sensors in the helmet. Users can hear sounds from headphones and interact with virtual objects using input devices like joysticks, wands, and data gloves. Using the Pubmed query ["virtual reality" and (training or learning)], the first paper among 3697, was indexed in 1991 and entitled "Virtual reality: a technology in nursing education's future." Two decades and half after, Li et al.[10] present a review of current virtual reality based simulators for medical application, especially in surgery training, pain management (VR used as a distraction therapy), and therapy for psychological diseases. Some of the most innovative projects using immersive technologies for medical education rely on Microsoft's mixed reality device, HoloLens. At Case Western Reserve, laboratories with cadavers and 2D illustrations in medical books are being replaced by HoloLens headsets. Using the HoloAnatomy app, medical students can rotate and virtually dissect a body to see the structures, systems, and organs. Using same tools, the Leiden University Medical Center in the Netherlands explores innovative ways to help students understand human anatomy. Medical students work with virtual anatomical models that mirror the movements of their own bodies, helping them understand anatomy from their own physical movements.[11] While attempting to narrow the query with [virtual reality" and (teaching or learning) and "medical education" and review], the most recent reviews of the 127 retrieved are mainly focused on surgical and endoscopy procedures: otorhinolaryngology[12-14] and otologic skills training[15], mastoidectomy,[16,17] neurosurgery,[18] ophthalmology[19] and intraocular surgery,[20] orthopedic[21] and spine surgery,[22] hysteroscopic,[23] endovascular,[24] and endoscopy training.[25-27] Only a few are related to the evaluation of effectiveness associated with immersive VR simulation. Stepan et al.[28] performed a randomized controlled study with 66 medical students involved in the instruction of neuroanatomy (33 in the control group and 33 in the experimental group experiencing immersive VR simulation). They found no significant difference in either the immediate postintervention or 2-month retention anatomy test scores between the control and VR groups. The groups had comparable baseline knowledge as indicated by the similar preintervention quiz results. Despite no significant improvement in retention, the VR group had a significantly better learning experience, reporting a higher level of learning content engagement, usefulness, and likelihood of endorsement to others. Authors concluded that a VR model of neuroanatomy was not less effective than traditional methods of teaching in terms of student exam performance. Peterson et al.[29] compared student examination performance on material taught using lecture and cadaveric dissection teaching tools alone or lecture and cadaveric dissection augmented with computerized 3D teaching tools. No correlation was observed between examination performance and number of times a student accessed the augmented materials outside of class. Student preference for the type of study material varied widely but did not correlate with examination performance.

In the context of medical education, the term **"virtual patient"** generally refers to any software that allows case-based training. The technical basis of VPs ranges from low-interactive Web pages to high-fidelity simulations or virtual reality scenarios. In the form of interactive patient scenarios, they are typically used to foster clinical reasoning skills acquisition in health care education.[30,31]. Interactive patient scenarios are Web-based applications in which a learner navigates through a VP scenario and interacts with the VP in form of menus, questions, or decision points. A variety of commercial and open-source VP systems, such as CASUS, OpenLabyrinth, or i-Human are available and applied in health care education.[32] Such systems provide tools for educators to create VP scenarios and deliver them to their students. Despite apparent consensus, Kononowicz et al.[30] have reported heterogeneity regarding the definition of VP among users. Thus, they developed a virtual patient classification framework that provides various communities (educators, VP developers, researchers in healthcare education) with a model based on predominant competency (knowledge, clinical reasoning skills, team training, procedural and basic skills, patient communication) and base technology (multimedia system, virtual world, dynamic simulation and mixed reality, manikin and part task trainer, conversational character) of the virtual patient. Moreover, a meta-analysis identified three main factors influencing the learner's perceived authenticity: authenticity of the VP story, the format of the VP, and the quality of the computer representation[33]. Thus Urresti-Gundlach et al.[34] published a coding framework based on patient data, patient representation, diagnoses, and setting, in order to make VP scenarios the most realistic and educational as possible. According to Hege et al.[35] Web-based VP systems must be improved while adding clinical reasoning into scenarios.

Interest for **"serious games"** by scientific community is easily measured by the production of articles. To date, the term "serious game" does not exist in the MeSH classification. Entry terms are: "game theory," "games, experimental," "games, recreational," or "video games." This last concept, introduced 20 years ago in 1996, is defined as: "a form of interactive entertainment in which the player controls electronically generated images that appear on a video display screen. This includes video games played on the home computers, and those played in arcades." It stands for computer games. In the MeSH tree hierarchy, this concept is a specialization of:

- Anthropology, Education, Sociology, and Social Phenomena Category > Human Activities > Leisure Activities > Recreation > Play and Playthings > Video Games
- Information Science Category > Information Science > Computing Methodologies > Software > Video Games

To date, the concept "video games" [mesh] indexes 4496 documents in Medline. The query ["video games" [mesh] and ("serious games" or "serious game" or "serious gaming")] finds 149 documents. It seems that the concepts "video games" and "serious games" do not cover the same objectives or the same objects. Based on what we defined before as "augmented reality," "virtual reality," and "virtual patients," we can consider that "serious games" are complex objects that combine them. It is now accepted[36,37] that a serious game is composed of new technologies (video games, 3D techniques, networks), depicts one or more elements of the game mechanics (pleasure, challenge, competition, desire to win, collaboration, strategy, reward), and is created with a "serious" intention that goes beyond mere entertainment to accomplish a specific goal. The player is the hero of the adventure. The design and use of the game is intended to induce a transformation in players such as the understanding of a phenomenon, a mechanism or a reasoning, the apprehension of a simple or a complex concept, the acquisition or enhancement of skills. Teachers say that the organization of a serious games tournament between students can also produce an emulation, develop the taste of the result, and the pleasure of learning to better understand.[38] Building a SG begins by creating a scenario (with or without rules, in other words a virtual patient) and by contextualizing the immersive environment (AR and/or VR). Regarding the instantiation of the "gameplay," that is to say the set of "fun rules" governing player(s) actions and interactions to achieve an objective, Alvarez et al. consider there are two types of gameplay[39]:

- the "Game" (gaming) type, which proposes precise objectives to be achieved and allows to evaluate the player's performance: win, loose, or a score.
- the "Toy" (playing) type, which does not propose explicit objectives to be achieved, and therefore does not judge/measure the player's performance. However, this does not prevent the player from setting goals himself.

In video games, the "rules" are rarely announced explicitly, but need to be identified mentally during the game. These "basic rules" can be translated into actions that constitute so many "gameplay bricks" which can return[39]:

- objectives to be achieved: to avoid (elements, obstacles, enemies, adversaries), to reach (to maintain one or more elements in a precise place or state of equilibrium), and to destroy (elements or enemies).
- means to achieve these objectives: to create (to assemble, to build, to create elements (graphic, sounds), to color, to draw from predefined patterns or brushes), to manage (to classify, to sort resources according to objectives to be achieved), to move (directing, driving an element or a character), to choose (predetermined choice among options, choice of an element, choice of a character or random, choice after generation of a random value), to fire (trigger a shot to reach one or more items located at a given distance), and to write (written or voice input as an answer or a trigger to launch a game functionality).

If certain intentions of serious games require relevant representation that awakes user's imagination[40] (in particular in children: comics, playful characters, simplified representations, or caricatures), in most other situations, especially for learning purposes, the reconstruction of realistic environments is an important element of the virtual contexts showed in serious games. For example, in Pulse/vHealthcare,[41] the clinical case creation framework in an emergency department, allows the choice of the characters (patients, professionals), the architecture and the medical furniture, allowing different locations in various countries (professional and socio-cultural realism).

In 2014, a national survey among French medical students,[42] as part of the evaluation of a serious game "Emergency Diagnosis,"[43] identified 3 dimensions students considered as the most important: validity of scientific content (89%), pedagogical scenario (86%), and realism of the scenery (85%). Another qualitative study about serious games used as training tools by companies[44] reports realism is defined as the level of congruence between business formalisms and culture of the company. This convergence is as much about vocabulary and graphics than procedures and responsibility for actions. Thus, a game featuring very cumbersome procedure, can stand in a fancy environment. On the opposite, a visually realistic game that does not take into account the reality of processes and knowledge will be ineffective.

In the field of medicine and healthcare, several authors propose to classify games according the specialty[45] (surgery, odontology, nursing, cardiology, first aid, dietetics and diabetes, psychology). Others have proposed to list games according to two axes[46]:

- "user" axis (professional and non-professional, research, academic, public health)

- "objectives" axis (prevention, therapeutics, evaluation, education, and informatics).

With the development of new games and the emergence of clinical studies evaluating the therapeutic impact, these initial approaches have been completed[37]:

- "objectives" axis (fun, health, skill acquisition);
- "functional" axis (cognitive or motor domain, interaction technology, 3D or non-3D interface technology, number of players, type of game, adaptability, progress monitoring, possibility of feedback, game portability, game engine, platform, connectivity);
- "step of the disease management process" axis (susceptibility, pre-symptomatic stage, clinical stage, post-treatment stage);
- "patient/non-patient" axis (disease monitoring, detection, treatment, rehabilitation, prevention, education, global health, well-being, training).

Wang et al.[47] proposed a systematic review of serious games in training health care professionals. Of the 42 serious games, 33 (79%) included a study design for evaluating the serious game as a teaching intervention. Of the 19 studies that attempted to evaluate their games for improving skill or knowledge gains, only 2 (11%) did not find significant differences between the intervention and comparison groups upon assessment or significant improvement after serious game use in 1 group pretest-posttest studies. Authors conclude that rigorous assessment is needed to establish best practices and to avoid serious games to lose credibility.

Speaking about digital innovations in simulation methods and tools, cannot be concluded without exploring "**digital debriefing.**" Debriefing after a simulation scenario is believed to be the single most important component of simulation-based education. In 2014, Levett-Jones et al.,[48] while reviewing different debriefing approaches, found one study that reported a statistically significant increase in outcomes for the groups exposed to video-facilitated instructor debriefing. Assuming computer-assisted video capture, we can consider, to some extent, that digital debriefing is feasible and can be relocated. Ahmed et al.[49] conducted a study with emergency medicine residents randomized into either a teledebriefing or on-site debriefing group during 11 simulation training sessions implemented for a 9-month period. The data revealed a difference between the effectiveness of traditional in-person debriefing and teledebriefing (as measured by the Debriefing Assessment for Simulation in Healthcare-Student Version). Teledebriefing was found to be rated lower than on-site debriefing but was still consistently effective.

As a conclusion, advances in technology have contributed to health care education. Technology-based interventions and simulations allow users to discover new contexts, to repeat risky procedures, and to evaluate their performance (knowledge, skills, velocity, precision) in safer and controlled learning sessions. Nevertheless, evidence has to be confirmed. Digital technologies remain tools whose role in the pedagogical and learning process must be refined. "Innovation" and "new technology" are frequently considered synonyms, although the hardware aspects of a new technology may be apparent, the social aspects and practical consequences of its software design and usage are typically much less obvious. As current technologies and their implementations will continue to evolve, the most appropriate use of educational technology in specific curricula must be continually sought. If our students are considered as "digital natives," they have to be taught to become "digital adaptive" as well as teachers. Furthermore, appropriate tools must be used to achieve relevant and reachable objectives. Digital exclusivity at any extreme position must be avoided. Mixed approaches seem to be the most effective (in terms of student performance) and best practices have to be found to practice (teach and learn) "digital augmented curricula."

References

1. Han F. Commencing a simulation-based curriculum in a medical school in China: independence and integration. In: *Healthcare simulation education: evidence, theory and practice.* Wiley Blackwell; 2017. p. 181–4.
2. Persson J. A review of the design and development processes of simulation for training in healthcare—a technology-centered versus a human-centered perspective. *Appl Ergon* 2017;**58**:314–26.
3. Moro C, Štromberga Z, Raikos A, Stirling A. The effectiveness of virtual and augmented reality in health sciences and medical anatomy. *Anat Sci Educ* 2017;**10**(6):549–59.
4. Lavallée S, Cinquin P, Szeliski R, Peria O, Hamadeh A, Champleboux G, et al. Building a hybrid patient's model for augmented reality in surgery: a registration problem. *Comput Biol Med* 1995;**25**(2):149–64.
5. Zhu E, Hadadgar A, Masiello I, Zary N. Augmented reality in healthcare education: an integrative review. *Peer J* 2014;**2**:e469.
6. Rolland JP, Wright DL, Kancherla AR. Towards a novel augmented-reality tool to visualize dynamic 3-D anatomy. *Stud Health Technol Inform* 1997;**39**:337–48.
7. Kugelmann D, Stratmann L, Nühlen N, Bork F, Hoffmann S, Samarbarksh G, et al. An augmented reality magic mirror as additive teaching device for gross anatomy. *Ann Anat* 2018;**215**:71–7.
8. *Augmented Reality Magic Mirror using the Kinect.* Available from: http://campar.in.tum.de/Chair/ProjectKinectMagicMirror; 2017 (last visit: November 26th 2017).
9. Barsom EZ, Graafland M, Schijven MP. Systematic review on the effectiveness of augmented reality applications in medical training. *Surg Endosc* 2016;**30**(10):4174–83.
10. Li L1, Yu F, Shi D, Shi J, Tian Z, Yang J, et al. Application of virtual reality technology in clinical medicine. *Am J Transl Res* 2017;**9**(9):3867–80.

11. Craig E, Georgieva M. *VR and AR: Driving a Revolution in Medical Education and Patient Care*. Available from: https://er.educause.edu/blogs/2017/8/vr-and-ar-driving-a- revolution-in-medical-education-and-patient-care; 2017 (last visit: November 26th 2017).

12. Sperry SM, O'Malley Jr BW, Weinstein GS. The University of Pennsylvania curriculum for training otorhinolaryngology residents in transoral robotic surgery. *ORL J Otorhinolaryngol Relat Spec* 2014;**76**(6):342–52.

13. Javia L, Sardesai MG. Physical models and virtual reality simulators in otolaryngology. *Otolaryngol Clin North Am* 2017;**50**(5):875–91.

14. Burns JA, Adkins LK, Dailey S, Klein AM. Simulators for laryngeal and airway surgery. *Otolaryngol Clin North Am* 2017;**50**(5):903–22.

15. Wiet GJ, Sørensen MS, Andersen SAW. Otologic skills training. *Otolaryngol Clin North Am* 2017;**50**(5):933–45.

16. Lui JT, Hoy MY. Evaluating the effect of virtual reality temporal bone simulation on mastoidectomy performance: a meta-analysis. *Otolaryngol Head Neck Surg* 2017;**156**(6):1018–24.

17. Andersen SA. Virtual reality simulation training of mastoidectomy—studies on novice performance. *Dan Med J* 2016;**63**(8) pii: B5277.

18. Konakondla S, Fong R, Schirmer CM. Simulation training in neurosurgery: advances in education and practice. *Adv Med Educ Pract* 2017;**8**:465–73.

19. Serna-Ojeda JC, Graue-Hernández EO, Guzmán-Salas PJ, Rodríguez-Loaiza JL. Simulation training in ophthalmology. *Gac Med Mex* 2017;**153**(1):111–5.

20. Thomsen ASS. Intraocular surgery—assessment and transfer of skills using a virtual-reality simulator. *Acta Ophthalmol* 2017;**95**(Suppl. A106):1–22.

21. Vaughan N, Dubey VN, Wainwright TW, Middleton RG. A review of virtual reality based training simulators for orthopaedic surgery. *Med Eng Phys* 2016;**38**(2):59–71.

22. Pfandler M, Lazarovici M, Stefan P, Wucherer P, Weigl M. Virtual reality-based simulators for spine surgery: a systematic review. *Spine J* 2017;**17**(9):1352–63.

23. Savran MM, Sørensen SM, Konge L, Tolsgaard MG, Bjerrum F. Training and assessment of hysteroscopic skills: a systematic review. *J Surg Educ* 2016;**73**(5):906–18.

24. See KW, Chui KH, Chan WH, Wong KC, Chan YC. Evidence for endovascular simulation training: a systematic review. *Eur J Vasc Endovasc Surg* 2016;**51**(3):441–51.

25. van der Wiel SE, Küttner Magalhães R, Rocha Gonçalves CR, Dinis-Ribeiro M, Bruno MJ, Koch AD. Simulator training in gastrointestinal endoscopy—from basic training to advanced endoscopic procedures. *Best Pract Res Clin Gastroenterol* 2016;**30**(3):375–87.

26. Harpham-Lockyer L, Laskaratos FM, Berlingieri P, Epstein O. Role of virtual reality simulation in endoscopy training. *World J Gastrointest Endosc* 2015;**7**(18):1287–94.

27. Naur TMH, Nilsson PM, Pietersen PI, Clementsen PF, Konge L. Simulation-based training in flexible bronchoscopy and endobronchial ultrasound-guided transbronchial needle aspiration (EBUS-TBNA): a systematic review. *Respiration* 2017;**93**(5):355–62.

28. Stepan K, Zeiger J, Hanchuk S, Del Signore A, Shrivastava R, Govindaraj S, et al. Immersive virtual reality as a teaching tool for neuroanatomy. *Int Forum Allergy Rhinol* 2017;**7**(10):1006–13.

29. Peterson DC, Mlynarczyk GS. Analysis of traditional versus three-dimensional augmented curriculum on anatomical learning outcome measures. *Anat Sci Educ* 2016;**9**(6):529–36.

30. Kononowicz AA, Zary N, Edelbring S, Corral J, Hege I. Virtual patients—what are we talking about? A framework to classify the meanings of the term in healthcare education. *BMC Med Educ* 2015;**15**:11.

31. Talbot TB, Sagae K, Bruce J, Rizzo A. Sorting out the virtual patient: how to exploit artificial intelligence, game technology and sound education practices to create engaging role-playing simulations. *Int J Gaming Comput Mediat Simul* 2012;**4**:1–19.

32. *Vpsystems. Virtual Patients (VPs) in Healthcare Education.* Available from: http://vpsystems.virtualpatients.net/[accessed 2017-11-26].
33. Cook DA, Erwin PJ, Triola MM. Computerized virtual patients in health professions education: a systematic review and meta-analysis. *Acad Med* 2010;**85**:1589–602.
34. Urresti-Gundlach M, Tolks D, Kiessling C, Wagner-Menghin M, Härtl A, Hege I. Do virtual patients prepare medical students for the real world? Development and application of a framework to compare a virtual patient collection with population data. *BMC Med Educ* 2017;**17**(1):174.
35. Hege I, Kononowicz AA, Adler M. A clinical reasoning tool for virtual patients: design-based research study. *JMIR Med Educ* 2017;**3**(2):e21.
36. Dörner R, Göbel S, Effelsberg W, Wiemeyer J. *Serious games foundations, concepts and practice.* Switzerland: Springer; 2016.
37. Wattanasoontorn V, Boada I, García R, Sbert M. Serious games for health. *Entertain Comput* 2013;**4**:231–47.
38. *Académie de Strasbourg. La notion de jeu sérieux.* Available from: https://www.ac-strasbourg.fr/pedagogie/ecogestion/numerique/jeux-serieux/#c9200 [accessed 2017-11-26].
39. Alvarez J. *Du jeu vidéo au serious game : approches culturelle, pragmatique et formelle. Thèse de Sciences, Université de Toulouse.* Available from: http://www.jeux-serieux.fr/wp-content/uploads/2008/04/TheseSeriousGames.pdf; 2007. [accessed 2017-11-26].
40. Stora M. Guérir par le virtuel. Presses de la Renaissance, 2005.
41. *Breakaway. Our games.* Available from: http://www.breakawaygames.com. [accessed 2017-11-26].
42. Jouan R, Pellet R, Staccini P. Evaluation d'un jeu appliqué à l'apprentissage en médecine. SeGaMed 2014, Nice.
43. *Ludomedic. Diagnostic en urgence.* Available from: http://www.ludomedic.com [accessed 2017-11-26].
44. *Ignition Factory. Serious game, l'âge de raison ?* Available from: http://ignition-factory.com/wp-content/uploads/2015/01/etude_serious_game.pdf; 2014 [accessed 2017-11-26].
45. Ricciardi F, Tommaso De Paolis L. A comprehensive review of serious games in health professions. *Int J Comput Games Technol* 2014; Available from: https://www.hindawi.com/journals/ijcgt/2014/787968/ [accessed 2017-11-26].
46. Sawyer B, Smith P. *Serious Games Taxonomy. Game Developers Conference.* Available from: https://thedigitalentertainmentalliance.files.wordpress.com/2011/08/serious-games-taxonomy.pdf; 2008 [accessed 2017-11-26].
47. Wang R, DeMaria Jr S, Goldberg A, Katz D. A systematic review of serious games in training health care professionals. *Simul Healthc* 2016;**11**(1):41–51.
48. Levett-Jones T, Lapkin S. A systematic review of the effectiveness of simulation debriefing in health professional education. *Nurse Educ Today* 2014;**34**(6):e58–63.
49. Ahmed RA, Atkinson SS, Gable B, Yee J, Gardner AK. Coaching from the sidelines: examining the impact of teledebriefing in simulation-based training. *Simul Healthc* 2016;**11**(5):334–9.

PART TWO

Digital Innovations - Design considerations

CHAPTER THREE

Designing digital education and training for health

Heather Wharrad, Richard Windle, Michael Taylor
School of Health Sciences, University of Nottingham, Nottingham, United Kingdom

Chapter outline

Toward a theoretical framework for designing digital education and training

Various learning theories are appropriate for the practical nature of healthcare professionals (HCPs) work to guide the design of digital education and training.[1] Those focused on social learning,[2] situated learning,[3] and experiential learning[4] are particularly relevant to vocational training. Three learning theories in particular provide a strong foundation for digital learning design. These are Carver's instantiation of experiential learning,[5] Wenger's community of practice,[6] and Mayer's multimedia learning theory.[7,8]

Digital Innovations in Healthcare Education and Training.
http://dx.doi.org/10.1016/B978-0-12-813144-2.00003-9

Carver's derivation of experiential learning for healthcare

Carver's[5] derivation of experiential learning theory is particularly useful as it contains four key principles that should underpin the design of HCP training and education.

These are: to ensure the **authenticity** of the material; to **actively involve** learners in the training; to ensure the **validation of their respective experiences,** on which the learning can be built; and to equip learners/trainees to **generalize their learning to new situations.**

Communities of practice

The value of Wenger's Communities of Practice theory,[6] in digital learning design, has been debated.[9] In participatory design where stakeholder views and expertise are harnessed in the production of digital learning, three elements are important. First, members of a "community of practice" should have a sense of connectedness and be able to effectively communicate as a group. Second, they should have an active and meaningful role within that community. The third characteristic of a community of practice is that members feel empowered. Participation as part of a community of practice aims to empower stakeholders to create effective learning design.

Multimedia learning theory

Mayer and Clarke[7] and Mayer[8] have published their research extensively on utilizing different media in digital learning. Based on the psychological theory around cognitive load, some of the key principles are shown in Table 3.1. People learn in different ways in different learning situations. By designing digital learning so that learners have a choice over which media they want to use to deliver the content will improve accessibility and the learning experience. So, strategies for some might include, for example, the ability to switch text on and off and just listen to audio while viewing an animation or video, for others switching off audio and reading the text might suit their preferences for learning and this may change in different learning situations and for different digital devices.

Instructional (pedagogical) design models

Within the overall theoretical principles afforded by the learning theories outlined in Section 3.1, a more practical framework is required to structure and link the elements of the digital learning resources. Common

Table 3.1 Some design principles for multimedia learning.

Principle	People learn better…..
Multimedia	..from words and pictures than from words alone.
Segmenting	..when a multimedia resource is presented in learner-paced segments rather than as a continuous unit.
Pre-training	..from a multimedia resource when they know the names and characteristics of the main concepts.
Modality	..from animation and narration than from animation and on-screen text.
Coherence	..when extraneous words, pictures, and sounds are excluded rather than included.
Redundancy	..from animation and narration than from animation, narration, and on on-screen text
Signaling	..when the words include cues about the organization of the presentation.
Spatial contiguity	..when corresponding words and pictures are presented near to each other on the page or screen.
Temporal contiguity	..when corresponding words and pictures are presented simultaneously rather than successively.
Personalization	..when the words are in conversational style rather than formal style.
Voice	..when words are spoken in a standard-accented human voice than in a machine voice or foreign-accented human voice.
Individual differences	Design effects are stronger for low-knowledge learners than for high-knowledge learners. Design effects are stronger for high-spatial learners than for low-spatial learners.

instructional design models include ADDIE[10] (Analyse, Design, Develop, Implement, Evaluate), Bloom's taxonomy,[11] and Gagné's[12] nine step model. By offering a rationale for the teaching and learning process at the level of the individual learner or groups of learners, such models help to ensure a consistency in approach and provide a reference-point for resolving challenges raised by collective critique of the training resources under development. Gagné's model[12] considers three important domains that impact on learning: affective, cognitive, and psychomotor, and so it is particularly suited to more vocational disciplines and it has frequently been used to design digital materials for teaching procedural or practical skills[13–21]. Laurillard's conversational framework[19] has been used substantively for digital learning design and categorizes the learning experience into five media forms—narrative, for example, video; interactive, for example, games and online activities; communicative, for example, web conference; adaptive, for example, computer simulation; productive, for example, animation. These classifications provide support and scaffolding for those involved in designing digital training to deploy the right technology based on user requirements and on pedagogic theory. Laurillard's framework[19] can be applied when designing digital learning such as RLOs where narrative, interactive, adaptive, and productive media elements are combined to produce an active and engaging learning experience for the user.

Reusable learning objects—an evidence-based digital e-resource design for healthcare training and education

One of the multimedia formats that can be seen to encompass a number of the principles that have been outlined earlier is the reusable learning object. Here we will explore what is meant by this terminology and how the different pedagogical principles are demonstrated, and the impact that this has had on effectiveness and sustainability.

Wiley[20] proposed the definition of an RLO to be "a digital resource that can be reused to facilitate learning." For some, an RLO is any digital resource—even a single digital image such as a photograph or an X-ray image. For others, it is a whole online module. Based on early research,[21] students wanted bite sized digital learning so from a learning design perspective, a useful definition describes a RLO as a web-based multimedia digital resource based on a single learning objective or goal, comprising a stand-alone collection of four components. **Presentation** of the concept, fact, process, principle, or procedure to be understood by the learner in order to support the learning goal. An **activity:** something the learner must do to engage with the content to understand it. A **self-assessment:** a way in which the learner can apply their understanding and test their mastery of the content. **Links and resources:** external resources to reinforce the taught concept and support the learning goal.[22]

By focusing on a specific learning goal or aim, RLOs can be self-contained and focused. RLOs should contain between 5 and 9 sections to optimize learning[23] with a learning time of 5–15 minutes. RLOs provide a visual and engaging way of enabling anyone to share their authentic stories, knowledge, and expertise. Whether this is a patient, healthcare professional, family member, student, or lecturer, all are experts and have a story to tell, thus validating the importance of their respective experiences as advocated by Carver's theory.[5]

Fig. 3.1 shows screen shots of pages of RLOs.

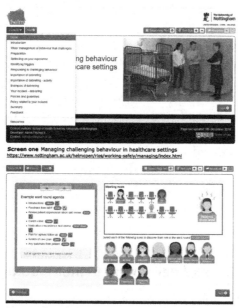

Screen one Managing challenging behaviour in healthcare settings
https://www.nottingham.ac.uk/helmopen/rlos/working-safely/managing/index.html

Screen two Ideal ward round
https://www.nottingham.ac.uk/helmopen/rlos/mentalhealth/ward-round/index.html

Screen three Identifying triggers for behaviour that challenges in healthcare settings
https://www.nottingham.ac.uk/helmopen/rlos/working-safely/identifying/index.html

Screen four Why use gloves?
https://www.nottingham.ac.uk/nursing/sonet/rlos/placs/gloves/index.html

Figure legend: *Screen one* is taken from the 'Managing challenging behaviour in healthcare settings' RLO and introduces the topic, and you can see the other sections of the RLO on the navigation bar.
Screen two contains a drag and drop activity from the 'Ideal ward round' RLO, and *Screen three* shows a short video based activity.
Screen four presents more information about appropriate and inappropriate glove use and having the self-assessment at the end of the RLO allows the learner to test themselves on what they've just learned.

Figure 3.1 *RLOs are short, sectioned, interactive, multimedia learning resources comprising between 5–9 sections.* This figure illustrates some of the sections of an RLO showing text, graphics, interactivities, and self assessment features. The template allows the learner to control various aspects of their learning experience including switching text and audio on or off, and controlling the media elements such as being able to pause or slow down the video or animation presentations.

The RLOs all have text and an audio commentary to deliver the main messages or learning content. Some users of the RLOs prefer to learn by reading text, others by listening to someone talking so the functionality to switch text or audio on or off is an important pedagogical feature. In addition, as advocated by Mayer's principles[7,8] (Table 3.1), the audio is available rather than having to read text when visual elements or animations are playing. So it's a good practice to offer both options. Laurillard's typology of technologies for learning[19] advocate the use of interactivities, questioning techniques, simulations, quizzes, and reflective exercises in contributing to engaging the learner and keeping them active in the process.

Using a design and development methodology

Why is a methodology important? One reason is that there are too many examples of poorly designed digital resources and training materials lacking a sound evidence base and the consideration of three key elements, participatory design, pedagogy, and peer review. This leads to rapid obsolescence and lack of widespread uptake and use.[24]

> *"Striving for a clear, simple and consistent conceptual model will increase the usability of a system."[25]*

A design and development methodology called "ASPIRE" uses participatory co-design principles and is centered on developing a community of practice approach to "unlock content" from experts and potential future users of RLOs. ASPIRE stands for aims, storyboard, populate/production, integration, release, and evaluate.[26,27] Each step is described in Fig. 3.2.

ASPIRE is underpinned by the principles emanating from the learning theories outlined in Section 3.1 and described in more detail in the following sections.

Theoretical basis for ASPIRE

ASPIRE contains steps common to a number of software development models[28] however it has been honed to fit optimally with requirements for designing high quality healthcare digital training. Software development models include a system or user requirements stage[29] at the start of the process, and in health, as in other fields, understanding of a particular concept, condition, or lifestyle is not confined to academics, but is situated within the groups of learners themselves whether they are healthcare students, patients, practitioners, charities, families, or support groups. One

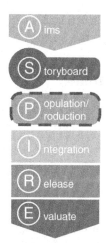

Figure 3.2 *The ASPIRE framework for designing digital training and education resources. A for Aims refers to the need to have a clear focus for the resource this includes the topic area to be covered or learning goal, and the characteristics of the target group of learners. S stands for Storyboarding, where stakeholders come together to work creatively to put together ideas for the content and design of the resource using storyboards. P for Populate and Produce is where the ideas are translated into media components ready to be Integrated—I together using a suitable platform such as HTML5. R for Release relates to how the resource will be made available to learners via a VLE, repository or website, for example, and how will it be promoted? E refers to the need to Evaluate the training resource to show its efficacy in a real learning situation.*

of the key challenges is to capture this unique knowledge in an accessible format, while one of the key opportunities is to provide a voice for these, often overlooked groups. In ASPIRE, a lot of emphasis is placed on involving end-users and other stakeholders at the start of the process but also at the quality review stages. This builds a community of practice around the development process and the final product giving users ownership, a voice and an acknowledgement of their expertise.

> *"This really brought something really novel to the process that we went through. So first off, we held a series of workshops and focus groups. We invited over 35 hearing aid users to come and take part. And what we were able to do with the workshop is to try and capture the personal perspectives, and thoughts, and attitudes around hearing loss and hearing aids and capture them all on large storyboards in picture format. The storyboards… …..really formed the content of the RLOs."*

However, in describing the ideation phase of the design thinking model, Neelan[30] makes the point that some creative ideas suggested by stakeholders may not necessarily enhance learning hence the need for pedagogists and learning technologists to guide the process.

Quality and evaluation

Quality

"Confidence in the learning quality of a resource can also be fostered by more robust attention to quality control to ensure the validity of the content and the pedagogical approaches adopted."[31]

A high level of quality in a digital training resource is achieved as a result of applying, collectively, a good standard in the following; conceptual and physical design, the accuracy and relevancy of content, usability, accessibility, formative and summative evaluation, and copyright and intellectual property.[32] A community of practice approach within a development methodology such as ASPIRE allow quality review at critical points in the process of designing digital training resources that can be overlooked when using more technology-driven methods.[33] The iterative stages of agile software development[28] involving regular review as the product develops are emulated in the ASPIRE process as a two stage peer review process normally carried out by one or two independent experts in the field. The first peer review involves a review of the written specification of the RLO prior to media development and focuses on accuracy of the content, appropriateness of the level for the type of learner, and alignment of the content and assessment to the learning goal of the RLO. The second review normally performed by the same reviewers occurs after the prototype RLO has been developed and is focused on how well the content has transferred from the written specification into the multimedia format.

If content authors (multiple stakeholders), working with developers in a participatory design model, produce consistently high quality resources, this is reflected in sustained use and reuse of digital resources.[33]

Evaluation

The use of self-reported "reactions" using rating scales to evaluate the quality of digital training interventions, categorized as level 1 according to Kirkpatrick's framework,[34] is commonly used because of the ease of collection of this level of data. However, the measurement of outcomes relating to learning gain (level 2), behavior change (level 3), and impact (level 4) will provide an increasingly comprehensive measure of efficacy of the digital training resources but are increasingly more difficult to achieve. The following case studies have used a range of evaluation tools from a validated RLO evaluation toolkit[35,36] (https://www.nottingham.ac.uk/helmopen/index. php/pages/view/toolkit) and other outcome measures relevant to the subject discipline to measure the impact of digital interventions for healthcare training in different target groups.

Case study 1: training healthcare students

People who live with a learning disability are an "underserved" community whose voices are sometimes overlooked and this is often reflected in the healthcare they receive. The SHOULD project (supporting health occupations understanding of learning disabilities)[37] aimed to address this, by enabling learning disability, nursing students themselves to design RLOs to inform healthcare students (including those not specializing in learning disability) on how to care for and work with these clients. The premise being that they are better placed to articulate approaches more likely to influence their peers. Some of the principles healthcare students should consider when working with people with a learning disability are to ensure the right to take part in community life, to experience valued relationships with all members of society, to make choices, both large and small, in one's life, to learn new skills and participate in meaningful activities with whatever assistance is required. Thirty-seven learning disability specialist healthcare students from seven higher education institutions from across the United Kingdom and Ireland were involved in the SHOULD project. Four RLOs were co-produced with these students using the ASPIRE process providing a voice for this underserved group (accessed at https://www.nottingham.ac.uk/helmopen/index.php/rlos/keyword/176).

"You can't be empowered and have a voice some of the time, you have to be empowered and have a voice all of the time." (Student participant in the RLO workshop.)

Case study 2: patient education

This case study demonstrates how RLOs have been effective in improving new hearing aid users' knowledge and confidence in using their hearing aids. Using the ASPIRE process, a series of RLOs were developed to address two main problems, first there was evidence that between 30% and 50% of new hearing aid users newly fitted with hearing aids did not use them.[39] Second, following audiology appointments, hearing aid users were not retaining information about how to use and adapt to their hearing aids. An expert panel of 33 hearing healthcare professionals, and workshops involving 32 hearing aid users and 11 audiologists contributed to the design and content development of the RLOs.[38] The workshops provided valuable analogies, metaphors, and anecdotes that could be incorporated into the RLO to explain concepts such as brain adaptation to new sounds, expectations when using a hearing aid for the first time, and communicate tactics. The participatory approach recognized these key stakeholders in the design process to create content for a user-friendly multimedia educational intervention, to supplement the clinical management of first-time hearing aid users. C2Hear was effective in a randomized controlled trial[39] and has had widespread global use including adaptation for American, and Chinese people. These RLOs can be found on Youtube at https://www.youtube.com/channel/UC_CO85ih5H68q5YSxMziidw.

Case study 3: healthcare professionals training

In this case, study the impact of RLOs on the knowledge, attitudes, confidence, and behavioral intention of registered children's nurses working with children and young people (CYP) admitted with self-harm was determined. The intervention consisted of a digital educational program composed of three RLOs (accessed at http://sonet.nottingham.ac.uk/rlos/mentalhealth/octoe/) that were co-produced with CYP service users, registered children's nurses, and academics. For those who completed the intervention, improvements in knowledge, attitudes, confidence, and clinical behavioral intention were shown. Qualitative findings suggested participants experienced: skill development; feelings of empowerment, and improved confidence; being more knowledgeable; being able to effectively communicate; reflecting on their own practice; and consideration of CYP emotional health and wellbeing in a broader context.[40]

Discussion

Just because a technology is innovative does not mean that its use to support learning and instruction will be innovative or effective.[41] The use of digital technologies for training and learning must be led by the pedagogy not the technology. As suggested by Graesser,[42] the development of theoretical models and tools is particularly important to estimate the quality of the designs of learning environments during or before their potential development to keep pace with digital innovation. The theoretical models offered by Carver's instantiation of experiential learning,[5] Wenger's community of practice,[6] and Mayer's multimedia learning theory[7,8] along with others[17,18,31,32] have provided important principles which should be considered in designing for digital learning.

The important role of multiple stakeholder communities in conceptualizing digital learning and its subsequent design has been discussed in this chapter. Network analysis studies where stakeholder groups articulate their roles and connections with others in the community while designing digital resources demonstrate this.[43,44] These studies also highlight the importance of the pedagogical design role in facilitating the process of digital learning design and also in the quality of the final product. Working in this way also requires a different way of working for the learning technologists and healthcare educationists and trainers who form part of these communities.[45,46] The emergence of the unique skill-sets, in professionals such as learning technologists and educators with expertise in pedagogical/instructional design,[47] are examples of changing staff roles within learning settings. This, "third space"[48] marks the merging of academic and professional roles as digital learning technologies, grow and mature in learning institutions. Digital learning technologies will continue to drive rapid changes in the ways that we teach and learn. New technologies such as AI, machine learning, and educational software aren't just changing the field for students, they are shaking up the role of educators, creating philosophical shifts in approaches for teaching and remodeling the classroom.[49]

The ever changing nature of the digital learning environment means that digital training resources, like RLOs, must adapt in order to remain accessible and relevant to the 21st century learner. Two of these issues are the changing nature of devices used to access the resources and changes in the technical development platforms. The gradual demise of Adobe Flash[50] presents a challenge for learning technologists in ensuring that online learning resources continue to be readily accessible by learners. Digital training resources that were created using older technologies such as Authorware, Shockwave, and Flash need to be upgraded. The use of multi-sized screen devices and the obsolescence of software such as Flash have led to a rethink of how to continue to deliver interactive, compatible digital resources. It is important to acknowledge the potential of mobile, personal and wireless devices to radically transform the way people learn.[51,52] Important considerations for optimizing cross-screen compatibility and mobile use without compromising quality content have led to the use of HTML5 as the main software tool for development together with intuitive front-end frameworks such as Twitter bootstrap, which enable the implementation of simple and flexible HTML, and Cascading style sheets (CSS) and JavaScript for popular scalable user interface components and interaction.

VR and AI applications for education and training are already being implemented.[53,54] AI advances are beginning to personalize learning with applications that are able to deliver different learning levels to suit the requirements of each student in a class student information that is gathered and analyzed will go beyond test scores to include monitoring facial expressions and social interactions.[55] Google, Apple, and Microsoft among others have set-up specialized programs to help push the implementation of new emerging technologies.[56,57] Apple, for example, have introduced Create ML a freely available open source development framework that promises an easy way for developers to implement machine learning models with no previous expertise required. While these new technologies present opportunities for educators and trainers to transform learning, there will be challenges in ensuring that the technologies are implemented in response to learning need and appropriate pedagogical design and learning theory.

References

1. Braungart MM, Braungart RG. Applying learning theories to healthcare practice. In: Bastable SB, editor. *Nurse as educator: principles of teaching and learning for nursing practice.* 3rd ed. Boston: Jones & Bartlett; 2008. p. 51–89.
2. Bandura A. *Social learning theory.* Englewood Cliffs NJ: Prentice-Hall; 1977.
3. Lave J, Wenger E. *Situated learning. Legitimate peripheral participation.* Cambridge: University of Cambridge Press; 1991.
4. Rogers C. *Freedom to learn.* 3rd ed. New York: Merrill; 1994.
5. Carver R. Theory for practice: a framework for thinking about experiential education. *J Exp Edu* 1996;**19**:8–13.
6. Wenger E. *Communities of practice: learning, meaning and identity.* Cambridge, UK: Cambridge University Press; 1998.
7. Clark RC, Mayer REE. *Learning and the science of instruction.* San Francisco: Jossey-Bass; 2003.
8. Mayer RE. *The Cambridge handbook of multimedia learning.* New York: Cambridge University Press; 2005.
9. Windle R, Wharrad H, McCormick D, Taylor M, Laverty H. Sharing and reuse in OER: experiences gained from open reusable learning objects in health. *J Interact Media Edu* 2010;**2010**(1) Art. 4, DOI: 10.5334/2010-4.
10. Peterson C. Bringing ADDIE to life: instructional design at its best. *J Edu Multimedia Hypermedia* 2003;**12**(3):227–41.
11. Airasian PW, Cruikshank KA, Mayer RE, Pintrich PR, Raths J, Wittrock MC. Anderson LW, Krathwohl DR, editors. *A taxonomy for learning, teaching, and assessing: a revision of Bloom's Taxonomy of Educational Objectives (Complete edition)* New York: Longman; 2001.
12. Gagné R. *The conditions of learning.* 4th ed. New York: Holt, Rinehard & Winston; 1985.
13. Woo WH. Using Gagné's instructional model in phlebotomy. *Edu Adv Med Edu Pract* 2016;**7**:511–6.

14. Belfield J. Using Gagné's theory to teach chest X-ray interpretation. *Clin Teach* 2010;**7**(1):5–8.
15. Khadjooi K, Rostami K, Ishaq S. How to use Gagné's model of instructional design in teaching psychomotor skills. *Gastroenterol Hepatol Bed Bench* 2011;**4**(3):116–9.
16. Buscombe C. Using Gagné's theory to teach procedural skills. *Clin Teach* 2013;**10**(5):302–7.
17. Ng JY. Combining Peyton's four-step approach and Gagné's instructional model in teaching slit-lamp examination. *Perspect Med Educ* 2014;3(6):480–5.
18. Laurillard D. *Rethinking university teaching: a conversational framework for the effective use of learning technologies*. London: Routledge Falmer; 2002.
19. Windle R, McCormick D, Dandrea J, Wharrad H. The characteristics of reusable learning objects that enhance learning: a case-study in health-science education. *B J Edu Technol* 2011;**42**:811–23.
20. Wiley D. Connecting learning objects to instructional design theory: a definition, a metaphor and a taxonomy. Wiley D, editor. *The Instructional use of learning objects*; 2001. [on-line] Available from: http://www.reusability.org/read/chapters/wiley.doc, (accessed 28 January 2007).
21. Wharrad H, Kent C, Allcock N, Wood B. A comparison of CAL with a conventional method of delivery of cell biology to undergraduate nursing students using an experimental design. *Nurse Edu Today* 2001;**21**:579–88.
22. Leeder D, McClachlan J, Rodrigues V, Stephens N, Wharrad HJ, McElduff P. UCeL: a virtual community of practice in health professional education. IADIS Web-based communities; 2004: 386–393.
23. Miller GA. The magical number seven, plus or minus two: some limits on our capacity for processing information. *Psychol. Rev.* 1956;**63**(2):81–97.
24. Raaff C, Glazebrook,C, Wharrad H. A systematic review of interactive multimedia interventions to promote children's communication with health professionals: implications for communicating with overweight children. *BMC Med Inform Decis Mak* 2014;14: 8. https://doi.org/10.1186/1472-6947-14-8.
25. Benyon D, Turner P, Turner S. *Designing interactive systems: people, activities, contexts, technologies*. Edinburg, United Kingdom; Pearson Education; 2005.
26. Windle R, Wharrad H, Coolin K, Taylor M. Collaborate to create: stakeholder participation in open content creation. In: *Association for learning technology conference (ALT-C) connect, collaborate, create*. University of Warwick; 2016.
27. Taylor M, Henderson J, Windle R, Wharrad H, Coolin K, Riley S. Collaboration in the heart of the MOOC. In: *Association for learning technology conference (ALT-C) connect, collaborate, create*. University of Warwick; 2016.
28. Munassar NMA, Govardhan A. A comparison between five models of software engineering. *Int J Comput Sci Issues* 2010; 7(5): 94–101.
29. Dhandapani S. Integration of user centered design and software development process. In: *IEEE 7th Annual Information Technology, Electronics and Mobile Communication Conference (IEMCON)*; 2016.
30. Neelan M. How to balance design thinking methodologies with evidence-informed learning design principles. Available from: https://oeb.global/oeb-insights/how-to-balance-design-thinking-methodologies-with-evidence-informed-learning-design-principles/; 2018.
31. Windle R, Wharrad HJ. Reusable learning objects in health care education. In: Bromage A, Clouder L, Gordon F, Thistlewaite J, editors. *Interprofessional e-learning and collaborative work: practices and technologies*. USA: IGI Global; 2010.
32. JISC Quality considerations 2014 [online]. Available from: https://www.jisc.ac.uk/guides/open-educational-resources/quality-considerations. [Accessed 15 January 2016].
33. Windle R, Wharrad H, Taylor M. Going global–views from the open educational landscape. *IATED. 12th International Technology, Education and Development Conference Valencia, Spain*. 5-7 March, 2018.

34. Kirkpatrick JD, Kirkpatrick WK. *Kirkpatrick's four levels of training evaluation*. USA: ATD Press; 2016.
35. Wharrad HJ, Morales R, Windle R, Bradley C. A toolkit for a multilayered, cross-institutional evaluation strategy. *World Conf Edu Multimedia Hypermedia Telecommun* 2008;**2008**:4921–5.
36. Morales R, Carmichael P, Wharrad H J, Bradley C, Windle R. Developing a multi-method evaluation strategy for reusable learning objects: an approach informed by Cultural-Historical Activity Theory. *1st European Practice-based and Practitioner Research Conference—Improving quality in teaching and learning: Developmental work and Implementation challenges*, University of Leuven, Belgium; 2006.
37. Windle R, Laverty H, Herman L, Hallewell B, Wharrad HJ. SHOULD: learning disability nursing students teach their peers. *Learn Disability Pract* 2010;**13**:26–9.
38. Ferguson M, Leighton P, Brandreth M, Wharrad H. Development of a multimedia educational programme for first-time hearing aid users: a participatory design. *Int J Audiol* 2018;**57**(9):1–10 DOI: 10.1080/14992027.2018.1457803.
39. Ferguson M, Brandreth M, Brassington W, Leighton P, Wharrad H. A randomized controlled trial to evaluate the benefits of a multimedia educational programme for first-time hearing aid users ear and hearing 2016;37: 123–136.
40. Manning JC, Latif A, Carter T, Cooper J, Horsley A, Armstrong M, et al. 'Our Care through Our Eyes': a mixed-methods, evaluative study of a service-user, co-produced education programme to improve inpatient care of children and young people admitted following self-harm. *BMJ Open* 2015;**5**(12):e009680.
41. Spector JM. *Conceptualizing the emerging field of smart learning environments smart learning environments 2014*;1:2. http://www.slejournal.com/content/1/1/2.
42. Graesser AC, Chipman P, King BG. Computer-mediated technologies. In: Spector M, Merrill D, van Merriënboer J, Driscoll M, editors. *Handbook of research on educational communications and technology*. New York: Taylor & Francis; 2008. p. 211–24.
43. Morales R, Carmichael P. Mapping academic collaboration networks: perspectives from the first year of the reusable learning objects CETL. *J Universal Comput Sci* 2007;**13**(7):1033–41.
44. Latif A, Windle R, Wharrad H. "I Can Use Things, but I Can't Make Anything": a qualitative exploration of team networks in the development and implementation of a new undergraduate e-compendium. *Res Learn Technol* 2016;**24**:32630. https://doi.org/10.3402/rlt.v24.32630.
45. Peacock S, Robertson A, William S, Clausen M. The role of learning technologists in supporting e-research. *Res Learn Technol* 2009;**17**:115–29.
46. Mitchell K, Simpson C, Adachi C. What's in a name: the ambiguity and complexity of technology enhanced learning roles. In: Partridge H, Davis K, Thomas J, editors. *Me, us, it! Proceedings ASCILITE2017: 34th international conference on innovation, practice and research in the use of educational technologies in tertiary education*; 2017, p. 147–151.
47. Ritzhaupt AD, Kumar S. Knowledge and skills needed by instructional designers in higher education. *Performance Improvement Quart* 2015;**28**:51–69.
48. Whitchurch C. Shifting identities and blurring boundaries: the emergence of third space professionals in UK higher education. *Higher Edu Quart* 2008;**62**:377–96.
49. Bernard Z. Here's how technology is shaping the future of education. Business insider UK TECH [Online]. Available from: http://uk.businessinsider.com/how-technology-is-shaping-the-future-of-education-2017-12?op=1&r=DE&IR=T; 2017 [Accessed 1st October 2018].
50. W3Techs. Flash Blog. Available from: https://w3techs.com/blog/category/cp-flash; 2016 (Accessed 28.2.19).
51. Traxler J. The current state of mobile learning. *Int Rev Res Open Distance Learning*; 2009: 8. www.irrodl.org/index.php/irrodl/article/view/346/875.

52. Payne KF, Wharrad H, Watts K. Smartphone and medical related App use among medical students and junior doctors in the United Kingdom (UK): a regional survey. *BMC Med Inform Decision Making* 2012;**12**:121.
53. Bonasio A. How VR and AI will supercharge learning. https://techtrends.tech/tech-trends/how-vr-and-ai-will-supercharge-learning/; 2018 [Accessed 1st October 2018].
54. Anon. What if technology could help improve conversations online? https://www.perspectiveapi.com/#/; 2018 [Accessed 1st October 2018].
55. Herold B. The future of big data and analytics in K-12 education. https://www.edweek.org/ew/articles/2016/01/13/the-future-of-big-data-and-analytics.html; 2016 [Accessed 1st October 2018].
56. Available from: https://developer.apple.com/xcode/[Accessed 1st October 2018].
57. Available from: https://developer.apple.com/machine-learning/ [Accessed 1st October 2018].

CHAPTER FOUR

The role of technical standards in healthcare education

Martin Komenda^a, Matěj Karolyi^a, Luke Woodham^b, Christos Vaitsis^c

[a]Institute of Biostatistics and Analyses, Faculty of Medicine, Masaryk University, Brno, Czech Republic
[b]St Georges's University of London, London, Great Britain
[c]Karolinska Institutet, Stockholm, Sweden

Chapter outline

Introduction

Improvements in medical and healthcare education are closely connected to proven concepts of education reforms, as emphasized by the statement on the Bologna process,[1] and also to a general use of methodological and technical standards. There are a variety of methodological reform efforts being advocated and pursued by researchers, educators, learning institutions, corporate trainers, and government leaders. Concepts such as student-centered learning, computer-based training, online learning, distance learning, just-in-time learning, and self-learning are widely accepted as having the potential to substantially improve the efficiency and effectiveness of learning. In general, all of these approaches need to be in compliance with one or more underlying technical standards[2] maximizing the quality, compatibility, and interoperability of a continuous progress in the education of medical and healthcare study programmes. The practical use of these technical standards goes hand in hand with systematic reforms of education where all stakeholders (i.e., faculty managements, academics, organizations

Digital Innovations in Healthcare Education and Training.
http://dx.doi.org/10.1016/B978-0-12-813144-2.00004-0

of health professionals and medical societies) have to be involved. Healthcare education and training standards represent a crucial part in the reforming process, which must usually overcome the following obstacles: (1) making medical and healthcare education and training more consistent and coherent across undergraduate and postgraduate study programmes, (2) reflecting general characteristics and features of an effective learning environment, (3) providing powerful tools for a transparent description of specific educational domains (i.e., curriculum development, learning objects, assessment, etc.), (4) setting up common rules, policies, and recommendations for education providers and involved stakeholders and (5) adopting and implementing standards–compliant systems advancing medical teaching and learning with the use of modern information and communication technologies.[3,4]

Today, various standards can be used in different domains of medical and healthcare education (e.g., sharing and reusing of learning objects, students assessment and evaluation, curriculum development blocks and relations, learner's profiles and behavior, quality assurance of courses, etc.).[5] From the perspective of higher education institutions, the integration of standards represents the only way to ensure both interoperability and reusability of educational objects and tools together, thus avoiding custom–made solutions and proprietary techniques and software. Professional organizations and societies standing behind standards–compliant systems emphasize a wide adoption of produced standards, which are usually well-documented and very easy to use in practice. These well-established international bodies produce and guarantee miscellaneous recommendations and policies connecting real teaching to international standardized frameworks. They promote an effective usage of modern information and communication technologies directly to institutions of higher medical education, which help to train young health professionals on a long-term basis. Professional organizations and communities from around the world, such as the Association of American Medical Colleges,[6] MedBiquitous,[7] the Association for Medical Education in Europe,[8] the European Federation for Medical Informatics Association,[9] the mEducator Best Practice Network,[10] the General Medical Council,[11] and the Medical Faculties Network[12] have created numerous guidelines and standards that appropriately link up the medical informatics domain with good healthcare practices.[3] The development of each individual standard meets the stakeholders' requirements and follows a strict process lifecycle under the supervision of a consensus body such as a steering committee. As an illustration, Fig. 4.1 shows how the MedBiquitous consortium creates a new standard, the American National Standards

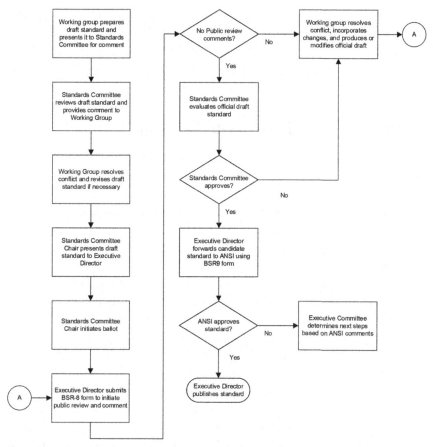

Figure 4.1 *Creation, approval, and publishing of a MedBiquitous standard (7).*

Institute approves it, and the standard is subsequently made public. Also included is a description, by role, of standards development responsibilities.[13]

This chapter reviews the usage of technical standards from several perspectives. Three of today's most up-to-date and discussed domains of medical and healthcare education are emphasized here: curriculum development and mapping, virtual patients and user behavior, and assessment of student learning. These domains are introduced by the following exploratory questions: (1) How do technical standards enhance the process of curriculum development and mapping in practice? (2) How can technical standards be applied in monitoring the behavior of users dealing with virtual patients? (3) How can the student performance be quantified and measured against student competency achievement by using technical standards, and how can be this employed as an indicator for improvements in the learning process?

(4) How can we measure the required uptake of theoretical knowledge and practical clinical skills of students by using a standardized and objective assessment process? We will answer these questions by investigating the role of contemporary technical approaches used in a standard-based infrastructure in medical and healthcare education.

Standards in curriculum development

Technical standardization in the curriculum domain addresses the need for structured data format describing individual curriculum building blocks together with their characteristics and relations. It usually serves for further processing, analysis, mapping, and visualization in order to have complex and transparent reports and statistics of a selected study program. The crucial motivation is to improve the traditional way of aligning medical and healthcare curricula and to strengthen the ability to identify all potentially overlapping areas, which are taught in clinical and theoretical courses. This agenda is performed by the involved stakeholders such as teachers, tutors, curriculum designers, institution managers, and others.[3] The following part summarizes the most relevant standards in terms of curriculum development, management, and mapping.

For the purposes of standards-compliant curriculum innovations, a set of proven data formats used in practice is required. One of the most productive association, the MedBiquitous Consortium, covers a wide-ranging group of medical and healthcare professional associations, universities, commercial organizations as well as governmental organizations. Among other outputs, MedBiquitous develops eXtensible Markup Language (XML) technical standards called the Competency framework (CF) and the Curriculum inventory (CI), which describe different pieces of a medical curriculum such as study programmes, sequence blocks, events, and competencies.[14] Based on a broad review,[4] the majority of today's curriculum management systems fully support at least one of the following standards:

- 4iQ Solutions platform (CI),
- Entrada (CI, CF),
- Ilios (CI, CF),
- LCMS+ (CI),
- MEDCIN (CI, CF),
- MedHub system (CI),
- MedSIS 3CKnowledge4You (CI),
- OASIS (CI),

- one45 (CI),
- OpenTUSK (CI),
- OPTIMED (CI, CF).

Based on this fact, we have chosen two of the earlier-mentioned Med-Biquitous standards, which can be used for curriculum description in the form of standardized entities. Other methodological and technical initiatives such as the CanMEDS Framework,[15] the Scottish Doctor Project,[16] and many more offer different alternatives for improving patient care by enhancing the training of physicians. In general, these approaches focus more on health practice meeting societal needs and residency training programmes rather than the education itself. From our perspective, the proper combination of CF and CI can provide a comprehensive guideline for linking the education content (represented by learning outcomes) with various building blocks (represented by medical disciplines, courses, and lectures).

The MEDCIN (Medical Curriculum Innovations) project[17] represents a pilot activity illustrating the conceptual way of an effective integration of these standards into education. It brings an innovative implementation of MedBiquitous standards as an example of best practices across Europe. MEDCIN aimed (1) to standardize medical and healthcare curricula; (2) to propose an innovative methodological background to map particular study programmes; (3) to develop and to implement a web-based platform for curriculum overview, analysis, mapping, and comparison purposes; all of these objectives were successfully achieved. Using the MEDCIN platform, stakeholders are able to parametrically describe and overview standards-compliant medical curricula and provide comprehensive information for a further verification by experts. Moreover, a set of analytical and data mining techniques for in-depth analyses are available here. It also proves the efficiency of curriculum standardization in practice on real institutional data. The MEDCIN database have been designed in accordance with MedBiquitous standards (CI, CF) and makes it possible: (1) to understand and to overview the structure of imported curriculum data, using a set of analytical and data mining techniques for computer-based evaluation and mapping of standardized medical curricula, (2) to compare individual curricula using a computational model for a systematic exploration of relations in a given curriculum based on data mining and natural language processing methods, (3) to promote the know-how of curriculum innovation, using a best-practice methodology for achieving the comparability of higher education institutions. Fig. 4.2 shows the relations between all mandatory components of the MEDCIN database schema.

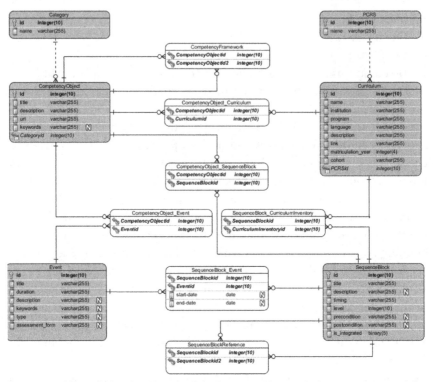

Figure 4.2 *The MEDCIN database model supporting the Curriculum inventory and the Competency framework.*

The orange colored entities represent the core building blocks of a curriculum (PCRS—Physician Competency Reference Set, Curriculum—study program, SequenceBlock—course or discipline, Event—lecture or seminar, CompetencyObject—learning outcome, Category—internal classification of learning outcomes). The white colored entities represent associative many-to-many relationships between selected core entities. Using this model, any curriculum compliant with the standards of CI or CF can be easily imported to the MEDCIN platform, which is able to eliminate poor transparencies in curricula and to improve the balance study programmes.

Standards in virtual patients and user behavior

It is increasingly recognized in medical education that clinical exposure on placements is not sufficient to provide the full breadth of experiences needed by learners to prepare themselves for clinical practice. These experiences are limited by a lack of access to and availability of rare and unusual cases or symptoms, since by definition these occur infrequently. In addition, patient safety concerns mean that only certain procedures or experiences are available to those in the relatively early stages of training. The use of simulation can help to supplement these experiences by providing access to a broader range of simulated conditions and diagnoses, and without placing patients at risk of harm. High-fidelity simulations, using mannequins or other technologies are widely used, and provide excellent opportunities for learners to receive feedback and work in groups or interprofessional settings. Such feedback is often formative for learning and development rather than being summative.[18] Low-fidelity simulations such as virtual patients (VPs) and case scenarios have also been explored as potential sources of sophisticated assessment, particularly when looking at the assessment of skills such as clinical reasoning, as opposed to simply testing knowledge.[19]

Let us assume that your institution provides a set of VPs accessible by any web browser. In general, it can be any online educational material—educational works, multimedia, presentations, books, etc. From the academic institution's perspective, monitoring of user behavior can provide very interesting information. Various phenomena can be tracked: (1) who are the users, (2) how much time the users spend on each activity, (3) when is a VP or other educational material used most frequently, (4) what the users search, (5) whether the users find desired results, (6) whether the users finished their commenced activities, and much more. The learning process could be optimized for individual learners if we were able to track all of the earlier-mentioned issues and to infer conclusions. The situation is even more complicated when information about user behavior is shared between multiple different e-learning systems.

Because the learning process of every person is unique, there is a need to track activities and share its content efficiently and in a standardized form. The Shareable Content Object Reference Model (SCORM) for e-learning is one of the currently existing sets of specifications and standards. It is produced by the advanced distributed learning (ADL) initiative.[20] SCORM is supposed to provide guidelines for the implementation of e-learning platforms and systems. If developers follow this purely technical standard, it leads to the creation of an interoperable learning software with no impact to instructional design or pedagogical habits.[21] In recent years, a new specification called the eXperience Application Programming Interface (xAPI) has been defined as a simpler successor of SCORM. It is prepared for a new era of e-learning in the form of learning on mobile devices, the usage of

simulators or teaching in virtual or augmented reality. xAPI is also prepared
for tracking users' activities in the real world and during the offline learning,[22] and information about a learner is collected in three main phases:

1. Learning—students learn from various materials.
2. Recording—each learning activity is recorded in a standardized form
   to the Learning Record Store (LRS). Each recorded statement fulfils a
   predefined form: Actor + Verb + Activity + Additional properties.
3. Sharing—activities recorded within the LRS are shared among various
   services and systems.

Analytics build earlier the xAPI provide a robust tool for the creation of
each learner's profiles. Thanks to the simplicity of statements, the standard
is very flexible and powerful. Full specification is available on GitHub pages
of the project.[23]

Standards in assessment

Although often only given consideration after other aspects of teaching and learning, assessment is a key aspect of any educational curriculum,
particularly when the endpoint of the curriculum is an academic award or
accreditation. Assessment is regularly criticized as having evolved far less
dramatically than other educational aspects over recent years. For reasons of
conservatism and caution, the traditional paradigm of assessment as being
learners completing closed-book written assessments in a formal, timed examination setting has often continued to be held as dominant. However, in
medical and healthcare education, this picture does not accurately represent
the broad range of assessment and assessment types currently in use. Medical
and healthcare practitioners are continually assessed from their very earli-
est days as an undergraduate, as they move into clinical practice, and then
throughout their career as they undergo a lifelong process of continuous
professional development (CPD). These assessments take many forms, some
of which are outlined later, along with the ways in which these have been
supported by technological enhancements.

Ultimately, assessment is a critical area for academic institutions, where
authentication, identification, reliability, and validity are key concerns for
ensuring that qualifications are not erroneously rewarded or denied based
on system failures. For this reason, many assessments are still carried out in
close to traditional forms—submission of essays, answering multiple-choice
questions, etc. Observational assessments that use technology rely on face-
to-face delivery still. Where technology has become involved is in helping

to facilitate these forms of assessment in a more efficient manner; the use of Electronic Management of Assessment (EMA) tools allows essays to be submitted online, and results to be shared and communicated digitally between the faculty and learners. Additional functionality is provided by tools such as TurnItIn,[24] which not only allow learners to submit their work digitally, but also make it possible for the faculty to monitor and to measure the incidents of plagiarism against existing online submissions.

Competencies

Assessments in medical and healthcare education can be knowledge-based, performance-based, practice-based, or behavior/attitudes-based,[25] but ultimately these assessments are all designed to be ones of different aspects of clinical competence.[26] Different competencies can be mapped to different types of educational activity to which they are well-aligned,[27] and often these will also prove to be useful assessment tools, whether for formative self-directed assessment or summative assessment for academic credit. The range of competencies against which medical and healthcare students can be assessed is vast, as evidenced by the dizzying array of learning objectives and outcomes that students at almost every academic institution are asked to measure their performance against. Indeed, there is very little consensus between institutions as to what these should be, with each institution often having their own list of competencies and outcomes available to learners through a system such as a Curriculum Management System (CMS). However, the MedBiquitous standards organization has developed a series of XML-based data standards to allow institutions or systems to describe their range of competencies. ANSI has ratified the MedBiquitous Competency Framework, ANSI/MEDBIQ CF.10.1-2012 as a verified data standard, which allows institutions to better describe competencies in a way which can be understood between different systems and institutions. This standard brings together additional specifications such as the MedBiquitous Competency Object and Curriculum Inventory ANSI/MEDBIQ CI.10.1-2013 in such a way that, by being universal and human-readable standards, they can underpin many of the other technical developments being described.

Objective structured clinical examination

A key assessment tool for many medical and healthcare disciplines is the Objective Structured Clinical Examination (OSCE), an approach which was first introduced in 1979[28] and has since gained widespread global acceptance and adoption as a means for assessing the range of clinical competencies expected by students. Numerous attempts have been made to deliver OSCEs with the use of online tools[29,30] primarily in a blended setting. However, there have been criticisms that there are signs of an increasing tendency for learners to "learn-to-test" in order to ensure that they pass OSCEs successfully, with advocates encouraging an increasing emphasis on continuous testing through practice-based assessments, workplace-based assessments, and e-Portfolios.[31].

Workplace-based assessments

Workplace-based assessment (WPBA) is widely recognized as being a key tool in providing feedback to learners[32]. Much workplace-based assessment is built around partly standardized measures such as the mini-Clinical Evaluation Exercise (mini-CEX), Case-based Discussion (CbD), and Direct Observation of Procedural Skills (DOPSs) as well as other forms. These assessments are generally used for formative learning and feedback, giving learners the opportunity to perform some form of clinical encounter or procedure, and then receive observed feedback from a more experienced member of clinical staff. However, they can also be used as part of a summative assessment process. There are numerous practical and administrative challenges that are associated with the use of WPBAs; the need to carry paper forms and to deliver those forms back to the home institution while on placement, the lack of validation on forms and illegibility of handwriting. Perhaps most importantly, the use of paper forms can encourage assessments to be completed by clinicians in bulk after the event has taken place, which can diminish the role of the face-to-face encounter and the verbal feedback which is inherent in the intended assessment design. Attempts have been made to develop technological solutions to address these challenges, particularly using mobile devices.[33] The use of electronic systems comes with additional challenges however, such as ensuring consistent network access at clinical sites if there is no offline data entry, as well as the training implications for both learners and the clinical assessors, who in many instances may not be affiliated to the academic institution in any way.

Examinations and tests

Of course, the assessment of knowledge is an essential part of training, and the use of single-best answers (SBAs) and multiple-choice questions (MCQs) have long been established as suitable tools for delivering reliable assessments. Widely used in both high-stakes summative and formative assessment, they provide the benefit of not requiring tutor or assessor marking;

no interpretation of responses is required, and the scoring is entirely objective and can be generated by a computer. By providing customized feedback for responses, such assessments can programmatically provide very personalized feedback for each learner. This makes them highly scalable to large numbers of learners, and suitable for embedding as assessments in a range of online tools, including Massive Open Online Courses (MOOCs). However, there has been criticism of this style of assessments due to a lack of sophistication. There is an element of chance that the learner can pick the correct answer from the list, which means that multiple questions that relate to the same knowledge are required in order to ensure that the assessment scores are reliable indicators of student abilities. Similarly, the presence of a list of possible responses can also have a "leading" effect on learners, giving them guidance and direction toward a correct answer that would not be present in clinical practice. This has led to the development of assessment styles such as very short answers (VSAs),[34] which remove the pre-determined list of responses, but still allow for scalable computer-based marking by making use of machine learning techniques to develop reliable marking algorithms.

Conclusions

This chapter presented selected medical education domains (curriculum development and mapping, virtual patient and user behavior, assessment of student learning) where technical standards can be smoothly used in order to show general theoretical backgrounds together with strong points and benefits of individual approaches. Firstly, the MEDCIN project promoting an innovative curriculum standardization by unification of existing theoretical methodologies and its powerful technological performance were introduced. The developed standards-compliant platform aspires to become an integral part of higher education reform efforts at the European level. Moreover, xAPI as a robust solution for user behavior tracking in virtual patient scenarios was briefly described in this chapter. xAPI provides effective collections of user performance data from any virtual learning environments in order to get clear and objective feedback. Finally, the role of standards in the measurement of theoretical and clinical competences, clinical communication skills, and student's feedback was also emphasized.

References

1. Patrício M, Harden RM. The Bologna process—A global vision for the future of medical education. *Med Teach* 2010;**32**(4):305–15.

2. Schoening JR. Education reform and its needs for technical standards. *Comput Stand Interfaces* 1998;**20**(2):159–64.
3. Komenda M. Towards a framework for medical curriculum mapping [Internet] [Doctoral thesis]. Masaryk University, Faculty of Informatics. Available from: http://is.muni.cz/th/98951/fi_d/?lang=cs; 2015 [cited 2016 Feb 8].
4. Vaitsis C, Spachos D, Karolyi M, Woodham L, Zary N, Bamidis P, et al. Standardization in medical education: review, collection and selection of standards to address. *MEFANET J* 2017;**5**(1):28–39.
5. Konstantinidis S, Kaldoudi E, Bamidis PD. Enabling content sharing in contemporary medical education: a review of technical standards. *J Inf Technol Healthc* 2009;**7**(6):363–75.
6. AAMC [Internet]. Available from: https://www.aamc.org/; 2015 [cited 2016 Feb 8].
7. MedBiquitous Consortium [Internet]. Available from: http://medbiq.org/. [cited 2015 Aug 17].
8. An International Association For Medical Education—AMEE [Internet]. Available from: https://amee.org/home. [cited 2019 Jul 9].
9. European Federation for Medical Informatics Association [Internet]. Available from: https://www.efmi.org/. [cited 2019 Jul 9].
10. mEducator | Multi-type content repurposing and sharing in medical education [Internet]. Available from: http://www.meducator.net/. [cited 2019 Jul 9].
11. The General Medical Council [Internet]. Available from: https://www.gmc-uk.org/. [cited 2019 Jul 9].
12. Schwarz D, Dušek L. The MEFANET project [Internet]. Available from: http://www.mefanet.cz/index-en.php. [cited 2011 Mar 18].
13. Ellaway RH, Albright S, Smothers V, Cameron T, Willett T. Curriculum inventory: modeling, sharing and comparing medical education programs. *Med Teach* 2014;**36**(3):208–15.
14. Ellaway R, Smothers V. ANSI/MEDBIQ CI. 10.1-2013 *Curriculum Inventory Specifications*, 2013.
15. Frank JR, Danoff D. The CanMEDS initiative: implementing an outcomes-based framework of physician competencies. *Med Teach* 2007;**29**(7):642–7.
16. Simpson JG, Furnace J, Crosby J, Cumming AD, Evans PA, David MFB, et al. The Scottish doctor–learning outcomes for the medical undergraduate in Scotland: a foundation for competent and reflective practitioners. *Med Teach* 2002;**24**(2):136–43.
17. MEDCIN project (Medical Curriculum Innovations) [Internet]. Available from: http://www.medcin-project.eu/index.php. [cited 2017 May 4].
18. McGaghie WC, Issenberg SB, Petrusa ER, Scalese RJ. A critical review of simulation-based medical education research: 2003-2009. *Med Educ* 2010;**44**(1):50–63.
19. Botezatu M. Virtual Patient Simulation: implementation and use in assessment. *Inst för lärande, informatik, management och etik/Dept of Learning, Informatics, Management and Ethics*; 2010.
20. ADL Initiative [Internet]. Available from: https://adlnet.gov/. [cited 2019 Jul 9].
21. SCORM Explained [Internet]. SCORM. Available from: https://scorm.com/scorm-explained/. [cited 2017 Dec 21].
22. Rustici Software. Experience API: Overview [Internet]. Available from: https://experienceapi.com/overview/. [cited 2017 Dec 22].
23. Contribute to adlnet/xAPI-Spec development by creating an account on GitHub [Internet]. Advanced *Distributed Learning (ADL)*. Available from: https://github.com/adlnet/xAPI-Spec; 2019 [cited 2019 Jul 9].
24. Promote Academic Integrity Improve Student Outcomes [Internet]. Available from: https://www.turnitin.com/. [cited 2019 Jul 9].

25. Ellaway R, Masters K. AMEE Guide 32: e-Learning in medical education part 1: learning, teaching and assessment. *Med Teach* 2008;**30**(5):455–73.
26. Wass V, Van der Vleuten C, Shatzer J, Jones R. Assessment of clinical competence. *Lancet* 2001;**357**(9260):945–9.
27. Cook DA, Triola MM. Virtual patients: a critical literature review and proposed next steps. *Med Educ* 2009;**43**(4):303–11.
28. Harden RM, Gleeson FA. Assessment of clinical competence using an objective structured clinical examination (OSCE). *Med Educ* 1979;**13**(1):39–54.
29. Hernandez C, Lewis A, Castiglioni A, Selim B, Cendan J. OSCE standards-setting procedure facilitated by digital technology. *Med Educ* 2013;**47**(11):1132–11132.
30. Meskell P, Burke E, Kropmans TJ, Byrne E, Setyonugroho W, Kennedy KM. Back to the future: an online OSCE Management Information System for nursing OSCEs. *Nurse Educ Today* 2015;**35**(11):1091–6.
31. Khan H. OSCEs are outdated: clinical skills assessment should be centred around workplace-based assessments (WPBAS) to put the 'art' back into medicine. *MedEdPublish* 2017;6 Available from: https://www.mededpublish.org/manuscripts/1284.
32. Norcini J, Burch V. Workplace-based assessment as an educational tool: AMEE Guide No 31. *Med Teach* 2007;**29**(9–10):855–71.
33. Lefroy J, Roberts N, Molyneux A, Bartlett M, Gay S, McKinley R. Utility of an app-based system to improve feedback following workplace-based assessment. *Int J Med Educ* 2017;**8**:207.
34. Sam AH, Hameed S, Harris J, Meeran K. Validity of very short answer versus single best answer questions for undergraduate assessment. *BMC Med Educ* 2016;**16**(1):266.

CHAPTER FIVE

Internet of Things in education

Stathis Th. Konstantinidis

University of Nottingham, Nottingham, United Kingdom

Chapter outline

Introduction

Healthcare education is continuously looking for new ways of delivering learning and teaching. Healthcare education was always an early adopter of new technologies, even before the Internet era. Video-tapes was used back at the beginning of 1980s in clinical teaching and in continuous professional development of psychiatrists,[1] while closed circuit television was used years before in order to transmit surgical operation to a medical meeting at John Hopkins.[2] Information and communication technology was adopted early on healthcare education.[3] E-learning spread and established quickly,[4,5] new teaching and learning methods[6] proposed and a range of technical standards[7] created in order to further enhance interoperability, discoverability, and reusability of resources.[8,9] In addition, different aspects of collaborative learning[10,11] proposed either by customized platforms or through social media.[12] Furthermore, the evolution in simulation as a

Digital Innovations in Healthcare Education and Training.
http://dx.doi.org/10.1016/B978-0-12-813144-2.00005-2

learning method continuous for many decades.[13] In medicine and health sciences problem based learning is a well-established practice that allows learners to train and acquire competences by combining existing knowledge and skills. An example of such training is the Virtual Patients that have been utilized for almost a decade.[14,15] An emerging with a rapid growth rate of 30% field is game-based learning. Game-based learning sees players as problem solvers and explorers. Based on the principles of problem-based learning, a game provides an environment for it.[16]

At the same time, a number of applications[17–20] and frameworks[21] proposed to use mobile devices in a context aware place. Recently the Internet of Things (IoT) came into the foreground envisioning a future in which digital and physical entities can be linked, by means of appropriate information and communication technologies, to enable a whole new class of applications and services.[22]

This chapter debates the role of Internet of Things in Education (IoTiE) in a healthcare context. Initially, a discussion and a definition of what considered as IoT will be given, followed by an overview of IoT in Healthcare. Next, the technologies used in IoT are briefly discussed, followed by an analysis of security issues and challenges for IoT. The theory underpin the IoTiE is explained, succeed by a discussion on the benefits and limitation of a context aware educational system interacting with environment or location-based learning games. The case of ViRLUS is presented as a representative combination of current and future digital innovations. Last but not least, the IoTiE concepts are summarized and future visions are discussed.

Defining the Internet of Things in education

While IoT is a trending buzzword, the notion of this smart, connected world exist for long time under different names. First Mark Weiser's described in an article on Scientific American in 1991,[23] his vision of ubiquitous computing, also called pervasive computing. The main facets of his visions were context aware computing; ambient/ubiquitous "intelligence"; and ambient tracking/monitoring of people and things. Satyanarayanan[24] 10 years later published his vision on pervasive computing stating that represents *"a major evolutionary step in a line of work dating back to the mid-1970s. Two distinct earlier steps in this evolution are distributed systems and mobile computing."* He also states that pervasive computing as a research area includes: effective use of smart spaces; invisibility; localized scalability; techniques for masking uneven conditioning of environments.

Kevin Ashton claims the first use of termIoT in 1999[25] linking the idea of RFID in a supply chain. Since then the term IoT evolved as the technology and the internet did. The oxford dictionary defines IoT as "*the interconnection via the Internet of computing devices embedded in everyday objects, enabling them to send and receive data.*" While Gubbi et al [26] state that "*although the definition of 'Things' has changed as technology evolved, the main goal of making a computer sense information without the aid of human intervention remains the same. A radical evolution of the current Internet into a Network of interconnected objects that not only harvests information from the environment (sensing) and interacts with the physical world (actuation / command / control), but also uses existing Internet standards to provide services for information transfer, analytics, applications, and communications.*"

The IEEE Internet Initiative definition as explained in Minerva et al[27] distinguish between low and high complexity. For low complexity systems the definition is: "*An IoT is a network that connects uniquely identifiable 'Things' to the Internet. The 'Things' have sensing / actuation and potential programmability capabilities. Through the exploitation of unique identification and sensing, information about the 'Thing' can be collected and the state of the 'Thing' can be changed from anywhere, anytime, by anything,*" while for complex systems the IEEE Internet Initiative devised the following definition of IoT: "*Internet of Things envisions a self-configuring, adaptive, complex network that interconnects 'things' to the Internet through the use of standard communication protocols. The interconnected things have physical or virtual representation in the digital world, sensing / actuation capability, a programmability feature and are uniquely identifiable. The representation contains information including the thing's identity, status, location or any other business, social or privately relevant information. The things offer services, with or without human intervention, through the exploitation of unique identification, data capture and communication, and actuation capability. The service is exploited through the use of intelligent interfaces and is made available anywhere, anytime, and for anything taking security into consideration.*"

Different definition can be explained if we consider that the IoT can been seen from different angles, either from an "Internet oriented" or a "Things oriented" perspective, while the "Semantic oriented" perspective should be also considered as the exchange of information become more challenging.[28]

IoT is interwoven with smartness notion transforming the way we are living, resulting to better quality of life, and increased process efficiency.[29] While there is a big debate around the definition of IoT, Cisco whitepaper[30] consider IoT evolution as internet of everything (IoE)—where everything is interconnected as a smart object and the line between the physical object and the digital information about that object is blur. According to Selinger et al.,[30] three key factors should be addressed for successful adoption of IoE in Education: Security, Data Integrity, and Educational Policies, while it identifies four pillars of IoE in education: People, Process, Data, and Things. The Person will be hyper-connected and becomes a "node" of the network, while persons will form new networks through discussion forums of MOOCs or other open global accessible resources. Persons big data analysis of everything in a learning analytics notion will be leading to more personalized learning and allow curricula to be restructured.[31] The Process allow the other three pillars to work together to "deliver value" in the connected world of IoE. Data being the third pillar will soon be send as higher-level information, instead of raw data, enabling data-driven decision making in education to enhance personalized learning, excite learners, and improve curriculum.

Based on the earlier, the definition of **Internet of Things in Education** (IoTiE) envisions a network of interconnected *objects* that not only harvests information from the environment (sensing) and interacts with the physical world (actuation/command/control), but also the reactions of the physical world shape the environment. To this extent, IoT uses existing Internet standards to provide services for information transfer, analytics, applications, and communications. The IoTiE is exploited through the use of intelligent interfaces and is made available anywhere, anytime.

Internet of Things in Healthcare

The average age of population continuously increases, complemented by an increased need of improved management of chronic diseases, causing the health systems to be under great pressure to deliver on time high quality care. IoT in Healthcare can be seen as alternative approach to personalized high-quality care, with its application covering the whole spectrum of care while service users are at home, on the go, or in a clinical setting. IoT in Healthcare can enhance health services to become more cost effective and allow better and personalized healthcare.[32,33]

Home healthcare can be more personalized and the home ecosystem can contribute to service users' care through sensors and Things at home. Vital signal monitoring (e.g., oxygen saturation, electrocardiogram, etc.), emergency situation management, rehabilitation system, medication management, wheelchair management, telemedicine are some of the potential applications that can be met at home.[34-36]

Healthcare on the go is widely accessible nowadays due to the wide spread of smartphones and the continuous connectivity through wireless networks. Phone sensors, wearable devices can be easily integrated with a smartphone, allowing continuously monitoring of health data, with a wide range of applications tailored to the needs of each service user, such as: self-management of chronic diseases, wellbeing and exercise, mental health, physiological monitoring (e.g., blood pressure, body temperature, etc.), instant communication with health professionals.[34,35,37]

Clinical care in different health settings can benefit in multiple ways form IoT applications to: direct care of service user, prevent hospital infections, equipment track and management, logistic management or even to reuse some of the smart cities applications for emergency situation management. Sensors, eatable devices, wearable devices, remote surgery robots can all contribute to clinical care.[34,35]

IoT in Healthcare is still at its infancy as a recent bibliometric analysis revealed,[38] since increased number of conference proceedings papers could be an indicator for the rapid expansion of the field and the need for presenting early results and concepts. It also identified eight trends in IoT in Health research: Systems/services design and implementation; Communication/connectivity protocols & algorithms; Industrial potential of IoT; Data science analysis, Storage, and connectivity; Quality management and privacy; Efficiency and cost of application; Smart cities; Ambient assistive living & active healthy aging. These research trends are also coupled by findings of other studies.[32,34,35,39,40]

Technologies used for Internet of Things
IoT architecture

Different IoT architectures have been proposed over the last years, following different models, such as the three-layer, the middle-ware based, the SOA based and the five-layer model.[41] The five-layer model is considered by many the most applicable to the IoT applications.[41,42] The five layers (Fig. 5.1) are:[41,42]

Perception Layer (or Device Layer or Objects Layer): Physical objects and sensors devices are forming this layer. Within this layer specific Things are identified and their specific information (e.g., motion, vibration, physiological signal, location, etc.) can be passed on to Network Layer.

Network Layer (or Transmission Layer or Object Abstraction Layer): This layer securely transfers the information from Perception Layer to Middleware Layer in order to be processed.

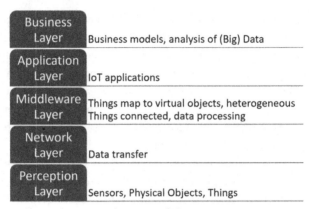

Figure 5.1 *5 Layers architecture of Internet of Things.*

Middleware Layer (or Service Management Layer): In this layer, Things and services are mapped to virtual representations, allowing heterogeneous Things to be connected and work together irrelevant of any hardware platform. In this layer, information is processed and automatic decisions are taken.

Application Layer: This layer provides the services requested by the users or applications (e.g., vital signs measurements, location, etc.). It provides high-quality smart services to meet user's needs.

Business Layer (or Management Layer): Within this layer, the overall management of the IoT system activities, applications, and services is taking place. Based on Application Layer data, business models, graphs, flowcharts, etc. can be build and can support decision-making processes based on Big Data analysis.

Standards and communication technologies for IoT

Communication technology is a core part of IoT in order to allow all the different "Things" to exchange vital information. Existing technologies can be used in different settings and based on different needs. The choice of the appropriate technology is influenced by multiple factors. Among others, these factors include short or long distances, whether the data are aimed to be transferred continuously or infrequent, the energy consumption and the security that they can provide. Table 5.1 provides a short overview of the most known communication technology with a short description and characteristics.

Table 5.1 Communication technology for Internet of Things.

Technology/Protocol	Description	Application field	Resource
Bluetooth Bluetooth low-energy (BLE) or Bluetooth Smart	A short-range communications technology	Wearable products; transfer small chunks of data; connectivity with smart phones	Standard: Bluetooth 5.0 core specification www.bluetooth.com/ specifications/bluetooth-core-specification
ZigBee	Low data-rates over a restricted area and within a 100 m range	Targeting applications that require relatively infrequent data exchanges	Standard: ZigBee 3.0 based on IEEE802.15.4 www.zigbee.org/zigbee-for-developers/zigbee/
Z–Wave	Low-power RF communications technology	Home automation; communication of small data packets	Standard: Z–Wave Alliance ZAD12837 / ITU-T G.9959 z-wavealliance.org/about_z-wave_technology/
6LowPAN (IPv6 Low-power wireless Personal Area Network).	6LowPAN is a network protocol that defines encapsulation and header compression mechanisms	Can be used across multiple communications platforms; designed for home or building automation	Standard: RFC6282 datatracker.ietf.org/wg/6lowpan/documents/
WiFi	Most common usage in homes today	Range of hundreds of megabit per second	Based on 802.11n
NFC (Near field communication)	Two-way interactions between electronic devices	Enables devices to share information at a distance that is less than 4 cm.	ISO/IEC 14443 A&B and JIS-X 6319-4 nfc-forum.org/nfc-and-the-internet-of-things/

(Continued)

Table 5.1 Communication technology for Internet of Things. (*Cont.*)

Technology/Protocol	Description	Application field	Resource
RFID (Radio frequency identification)	Automatic identification method, relying on storing and remotely retrieving data using devices called RFID tags	RFID tag is a small object that can be attached to or incorporated into a product. RFID tags contain silicon chips to enable them to receive and respond to queries from an RFID reader/writer.	ISO/IEC 15,693
Infrared data association (IrDA)	Short range (< 1 m), line-of-sight; communication standard for exchange of data over infrared light	IrDA interfaces are frequently used in computers and mobile phones.	Set of protocols for wireless infrared communications www.irda.org
Thread	IP-based IPv6 networking protocol	Designed as a complement to WiFi; capable of handling up to 250 nodes with high levels of authentication and encryption	Standard: Thread, based on IEEE802.15.4 and 6LowPAN www.threadgroup.org
Cellular	GSM / 3G / 4G cellular communication capabilities	High quantities of data; Expensive; power consumption	Standard: GSM/GPRS/EDGE (2G), UMTS/HSPA (3G), LTE (4G)
Sigfox	A wide-range technology uses the ISM band low data-transfer speeds	M2M applications that run on a small battery and only require low levels of data transfer	Standard: SigfoxA www.sigfox.com/en
Narrowband IoT (NB-IOT)	A narrowband radio technology; a low power wide area network	Improved indoor coverage, support of massive number of low throughput devices, low delay sensitivity, ultra-low device cost, low device power consumption and optimized network architecture.	www.3gpp.org/news-events/3gpp-news/1733-niot Specification Release 13

LoRaWAN	Wide-area network (WAN) applications low-power WANs	Optimized for low-power consumption and supporting large networks with millions of devices; secure bi-directional communication in IoT, M2M and smart city and industrial applications.	Standard: LoRaWAN lora-alliance.org/about-lorawan
Worldwide interoperability for microwave access (WiMax)	Can operate at higher bitrates or over longer distances but not both	Providing; portable mobile broadband connectivity across cities and countries through various devices; a wireless alternative to cable and digital subscriber line (DSL) for "last mile" broadband access; Smart grids and metering	Standard: IEEE 802.16 http://wimaxforum.org/home
LR–WPAN	IEEE Standard for low-rate wireless networks	Small sized packages; Low power consumption; 40—250 Kb/s; Range: 10–20 m	Standard: IEEE 802.15.4 ieeexplore.ieee.org/document/7964803/
MQTT	A machine-to-machine (M2M)/"Internet of Things" connectivity protocol.	Extremely simple and lightweight messaging protocol, designed for constrained devices and low-bandwidth, high-latency or unreliable networks	MQTT v3.1.1 is an OASIS Standard www.mqtt.org
Hypercat	A hypermedia catalog format designed for exposing information about the IoT assets over the web	Aims to solve the problem of a lack of interoperability in IoT devices that is preventing the exponential growth in the number and combination of IoT clients and servers	Standards: PAS 212:2016 www.hypercat.io/ www.bsigroup.com/en-GB/industries-and-sectors/Internet-of-Things/

(Continued)

Table 5.1 Communication technology for Internet of Things. (*Cont.*)

Technology/Protocol	Description	Application field	Resource
Very simple control protocol (VSCP)	An open and free framework/protocol for IoT/M2M automation tasks	A free and open solution for: • Device discovery and identification. • Device configuration. • Autonomous device functionality. • Secure update of device firmware • A solution from sensor to UI	Specification:VSCP Specification 1.10.18 – 2018-03-20 www.vscp.org
Network functions virtualization (NFV)	A network architecture that virtualize entire classes of network node functions	Provides a way to reduce cost and accelerate service deployment for network operators by decoupling functions like a firewall or encryption from dedicated hardware and moving them to virtual servers; Enables near real-time analytics and business intelligence	Standards: ETSI GR NFV-IFA 022V3.1.1 (2018-04) ETSI GS NFV-SEC 014V3.1.1 (2018-04) www.etsi.org/technologies-clusters/technologies/nfv

Furthermore, a number of organizations have released standards and recommendations for IoT including CEN/ISO, CENELEC/IEC, ETSI, IEEE, ISO, IETF, ITU-T, OASIS, OGC, and W3C.[43] For example, IEEE working on a "Standard for an Architectural Framework for the Internet of Things (IoT)"[a] and on a "Standard for a Reference Architecture for Smart City (RASC)"[b] aiming to promote cross-domain interaction, interoperability, and functional compatibility. While, the TU's Telecommunication Standardization Sector (ITU-T) have a range of working groups and initiatives working on and proposing regulations and standards for many IoT areas.[c] A recent overview and a review of IoT standards can be found in Guillemin et al.[43] and Trappey et al.,[44] respectively.

Security challenges for Internet of Things

While security continuously is a hot topic in the new technological era, IoT boost this further as it enables communication between heterogeneous entities and networks. IoT include communication between humans, humans and Things, Things and Things,[45] and increases the complexity of the network.

You cannot always have direct control with whom or what the Things are communicating, due to different types of constraints and heterogeneous communication, bootstrapping of a security domain or operations.[45,46] IoT security focus not only on the security services, but in the overall system,[45] thus multiple challenges arise on IoT. The following table (Table 5.2) depicts an overview of security challenges for IoTiE.[47–53]

Furthermore, users' actions to enhance their security should be prioritized. "Things" can be easily identified nowadays through relevant search engines such as Shodan (https://www.shodan.io)[55] or Censys (https://censys.io)[56]; thus assumptions that Things are accessible only by entities given the hostname and port number are dangerous.[51]

While already a number of standards, frameworks, protocols, and best practices have already been proposed, there are many technical issues to be solved, but also enabling the users to take action by raising their awareness.

[a] https://standards.ieee.org/develop/project/2413.html
[b] https://standards.ieee.org/develop/project/2413.1.html
[c] https://www.itu.int/en/ITU-T/techwatch/Pages/internetofthings.aspx

Table 5.2 Overall security challenges for IoT in education.

Focus area	Definition	Challenge
Confidentiality	Ensure that data are accessed by authorized users (Humans and Things) only	The cryptographic security is a challenge, balancing between lightweight cryptographic algorithm and higher performance of sensor node. In addition, a regulatory frameworks and security laws are needed
Authentication	Communication established with the desired Human or Thing only	Unauthorized authentication might result to unauthorized access to servers, systems, sensors, and networks. That would lead not only to access of sensitive users' information, but also to access of sensors/things, sending misleading information that might be dangerous for the users (e.g., lower the temperature of a fridge room where no people supposed to be, etc.)
Access	Permission in accessing data, communication infrastructures and resources.	The entirely system might become insecure, if access to the administrator privileges happens. Examples include a denial of service attack, which can be easily done through the communication with another network or controlling of Things in the space that might result in a failed learning journey.
Privacy	Determination of the data that can be shared with third parties	Sensitive user data, sensitive patient data (if real data are used), habits, interactions, health service private information might be leaked while tutors and organizations rumor negatively influenced.
Trust	A trustworthy evaluation between Things and/ or people interaction can define trust. Things in IoT can have social relationships.	What Things are trusted is not always known as it is not known in advance, what Things are going to be added in the IoT. Thus, the IoT network might be vulnerable if a not trusted Thing connects and exchange information with other Things and humans.

Table 5.2 Overall security challenges for IoT in education. (*Cont.*)

Focus area	Definition	Challenge
Middleware security	The heterogeneity of the system due to many different types of Things and persons connected, needs one or more middleware layers to be added in order to enhance security	Use of multiple middleware layers to enhance security might result to problems related to interoperability, task allocation, security, privacy, trust, and network performance among others.
Mobile security	Mobile Things or Humans are moving from one cluster to other and cryptography based protocols have to provide identification, authentication and privacy protection fast enough.	Users might result in no connection, while moving, or the system might connect users that do not have the authorization to do so, resulting to data breaches.
Regulatory frameworks	All the applicable regulatory framework, policies and laws.	According to Weber[54] (2015) the challenges are: Determination of the relevant types of privacy infringements; quality of data and quality of content; Identification of transparency and data minimization requirements; acknowledgement of interoperability and connectivity

Theoretical underpin
Educational theories

IoTiE can make the learning experience more authentic and meaningful. IoTiE as an educational intervention builds upon the Situated learning theory.[57] According to this theory, learning is a function of the activity, context, and culture in which situated. Thus, learners become part of the "community of practice,"[58] embodying certain beliefs and behaviors, which enable them to engage in stages and assume the role of the expert, following the process of "legitimate peripheral participation."[59] Drawing on social learning theories Bandura's[60]and Vygotsky's,[61] social interaction can lead to mitigation, help them to develop a deeper understanding. In addition, "learning by doing"[62,63] is also utilized within IoTiE as the theoretical knowledge applied in practice in any context.

Furthermore, Cognitive flexibility theory[64] promotes knowledge acquisition and application in ill-structured domains. Following Cognitive flexibility, IoTiE is able "*to restructure one's knowledge spontaneously, in many ways and in an adaptive fashion. Function of the way knowledge is represented (along multiple rather than single conceptual dimensions) and the process that operate on those mental representations (schema assembly rather than retrieval)*."[64] Learning through IoTiE applications implies the use and the enhancement of the "*ability to adaptively re-assemble diverse elements of knowledge to fit the particular needs of a given understanding or problem-solving situation*."[64]

IoTiE encompasses the concepts of Problem Based Learning (PBL) and as such activate the student prior knowledge, elaborate prior knowledge to activate the processing of new information, enable contribution of group discussion and social learning, restructure knowledge to fit the problem, and the epistemic curiosity can be expected to emerge.[65] At the same time the eight principles proposed in[66] can also inform IoTiE within a healthcare context: "*adult learners are independent and self-directed; adult learners are goal oriented and internally motivated; learning is most effective when it is applicable to practice; cognitive processes support learning; learning is active and requires active engagement; interaction between learners supports learning; activation of prior knowledge and experience supports learning; and elaboration and reflection supports learning*."

The role of debriefing at the end of a scenario based learning[67,68] or problem-based learning approach[69] is vital, as it can enhance the knowledge of the participants[70,71] being the last stage of experiential learning. Similarly, in IoTiE, a debriefing session or a self-reflection[72] for mature learners can enhance the acquisitions of skills and competences.

Behavioral triggers and motivational framework

Persuasion can be summed up in the following buzzwords: behavioral change, attitude change, motivation, change in worldview and compliance,[73] and similarly, persuasion defined as the deliberate use of communication to change attitudes and behavior of people.[74] To this extent, "Captology" an acronym based on the phrase: "Computers as Persuasive Technologies" is a term that coined by Fogg,[73] "*focuses on the design, research and analysis of interactive computing products created for the purpose of changing people's attitudes or behaviours."* Persuasive technologies are part of people's everyday life[75] and they should not be left out from serious games, since in specific context persuasive technologies can play a positive role by convincing stimulating or motivating users to engage in health and learning behaviours.[76]

Designing and implementing persuasive technologies is not an easy task. According to Kort et al.[77] many failures might occur before creating a successful product. Oinas-Kukkonen and Harjumaa proposed the Persuasive System Design (PSD) model, a framework for designing and evaluating persuasive systems, in which 28 design guidelines were defined.[78] Those strategies could be fulfilled by "personalizing," a strategy proposed by King and Tester[75] as a general strategy. Hawkins et al.[79] divide tailoring strategies into personalization, feedback and content matching. Though the strategies are usually combined, they do constitute unique conceptual ideas. Personalization attempts to increase attention or motivation to process the conveyed messages through for instance techniques like "identification," "raising expectations," or "contextualization." Feedback presents to the individuals their own or their competitors/collaborators health status or their activities. Content matching tries to give the user relevant information on the behavior of interest.

Pervasive, context-aware or location-based games for education

Pervasive games or location-based games (LBG) with a pedagogical aim can be considered a form of IoTiE, as its pervasive nature fits well with some of the definition for IoT. Location-based games are games that the users tasks and activities triggered by the location of the user. In the digital age, the most common device for user to use is a smartphone and location retrieved through smartphone's GPS, while a mixture of virtual and real-world data are used connecting virtual world to real life.[80]

LBG have been used in some educational programs,[81] and can be met in different types,[82] such as location based storytelling,[83] action games.[84,85] treasure hunts,[86,87] or role playing game.[88,89] On the other hand, many commercial LBG have been deployed with great success over the last years and a lot of them can be found in the "Encyclopedia of location-based games (or GPS-games)."[d] The usual infrastructure of such games consist of[82]: (1) the game engine, (2) the virtual space, and (3) the user profile database.

[d] https://dasbox.be/encyclopedia-of-location-based-games/

LBG in healthcare have been initiated for quite a few years now, with the main application to be physical training—exergames (games to enhance physical activities).[90] While gamification is widely adopted in healthcare education,[91] examples on LBG are difficult to be found[92]; while in the broader area of digital innovations in healthcare education, the following game can be considered. Seniors are learning for technology with the help of young volunteers utilizing tablets and mobile technology.[93] The target for each senior-young team was to find all locations, close the wormholes and restore the contemporary buildings, while the results of the effectiveness of the game were promising.

The case of the ViRLUS—a virtual reality learning ubiquitous space for "real" games

In this subsection, the case of ViRLUS—virtual reality learning ubiquitous space—will allow to get an insight of an IoTiE application, and its elements.

ViRLUS is modular system aims to transfer a mixed reality game technology to both educational and healthcare contexts, by not only utilizing IoT in games as an education tool, but also measure and realign the application effectiveness, service users engagement, creativity and collaborative behavior considering the potential risks and challenges that may arouse.

The ViRLUS allows to identify/validate how mobile technology enables a virtual reality learning experience within a real environment in the era of IoT and serious games. The ViRLUS games/cases fed by their users, but also from the environment itself. Thus, patients which are visiting a hospital will be able to record their "journey" into the hospital along with their treatment and at the same time play a game to be informed about procedures and health issues that they are relevant to them; health/medical students that they are in their practice will be able to play a game by interacting with a Virtual Patient in the real space and at the same time record the professional's side of patients "journey"; the environment itself (health unit) provides information to the game case by monitoring healthcare tools through sensors and by the provision of information through identifiers (RFID, RTLS—Real Time Location Systems or GPS locators). ViRLUS considers, in depth, the learning and behavioral triggers for all the service users as well as the social science aspects that arouse in healthcare environment context. In addition, ViRLUS gamified application targeting patients/service users visiting a ViRLUS registered healthcare provider for the provision of health-trusted information in a gamified and personalized mean.

ViRLUS has three main components:

ViRLUS game: It is a suit of different game-based scenarios experience in a real world interacting with Virtual Patients and situations. It is composed of *location tracking* for people and objects module, a *game engine*, a *tailored interface*, an *editor* to insert described game-based scenarios into the game, tracking of *activity data,* and providing feedback to the learner through the *analytics module*, and the *debriefing and performance component.*

ViRLUS gamified app: It is a gamified application targeting patients/ service users visiting a ViRLUS registered healthcare provider for the provision of health-trusted information in a gamified and personalized mean. Patients get the knowledge and following behavior triggers to increase their motivation to become health literal in the subject area, they are interested in. This module makes use of location tracking and record patient/service user interaction with the environment. In addition, this module gets users feedback for their experience in the healthcare setting and if the users prefer they can provide a patient summary to the system.

ViRLUS Game Based Scenario Interpreter: This ViRLUS module provides automatic new game—based scenarios. Thus, is the main vehicle to create new games at a low or no cost. As inputs modeled data through ontological representation are fed to the system including learners' *activity data, Patient's Scenarios Data, Environment Object Location Tracing, Environment Representation,* and *Existing case based scenarios (e.g., from Virtual Patients).* The output of the system is either a new scenario or an updated one based on real input. The Game Based Scenario Interpreter Module (GBSIM) reads and parses input data resources, related included information using semantic reasoning algorithms and, finally, produces a game-based scenario. The Game Based Scenario Interpreter has three components, which are described as follows:

Input data parser component

The following figure (Fig. 5.2) provides an overview of the input and outcome of the module. Based on an Existing scenario (e.g., VP scenario extended with localization data), different paths in the game are created based on different input (color and letters in the output game-based scenario). Existing states may get a location from the environment representation. Thus a lot of scenarios can automatically be created by interacting with environment (users and objects).

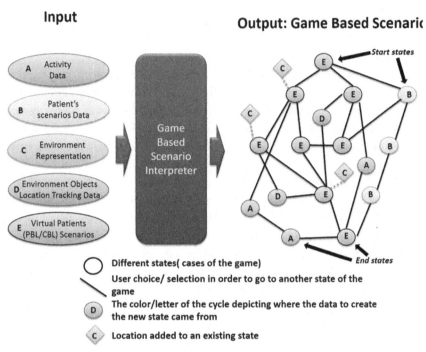

Figure 5.2 *Overview of the input and output of the game based scenario interpreter.*

Data indexing and relationship component

This component is responsible for indexing information, using a semantic schema based on used ontologies contained in: Existing scenarios (e.g., VP scenarios extended with localization data); Activity Data; Patient Scenarios Data; Environment Objects Location Tracking Data; and Environment Representations. The component produces relationships with the elements of an existing virtual patient (VP), enriching its scenario. The VP must follow the ANSI/MedBiquitous Virtual Patient standard package format. Within each VP, there are a series of linked pages or "nodes" that define the options available, each of which can be enhanced with a number of behaviors and services to further structure your experience and gameplay. The component can enrich the node's information, as well as creating new nodes according to the input data. Rules are functions attached to nodes that change the way a labyrinth (a case) is presented to the learner. One set of rules might set or change one or more counters (e.g., score) while another could require to have visited other nodes before being able to load the current one. The component can create rules, or change existing rules.

The game-based scenario export component

Finally, this component is responsible to export a structure that will represent the game-based scenario. The export result is based on used ontologies and can be modified or extended easily, or use it as is:

- **ViRLUS Silos:** Are the interconnected silos of mEducator and HELM Open which host health trusted educational resources, case based learning scenarios, and (location) game-based scenarios. They feed the ViRLUS game on request by the use of the Communication API.
- **Communication API:** is responsible for the communication between the different modules of the system. It utilizes the different ontologies used in ViRLUS.

   The following figure (Fig. 5.3) depicts an overview of ViRLUS system.

   Persuasive techniques used into the ViRLUS system as both behavioral and motivational elements. Among the several attributes of persuasive technologies proposed in,[73] three key attributes are used in ViRLUS:

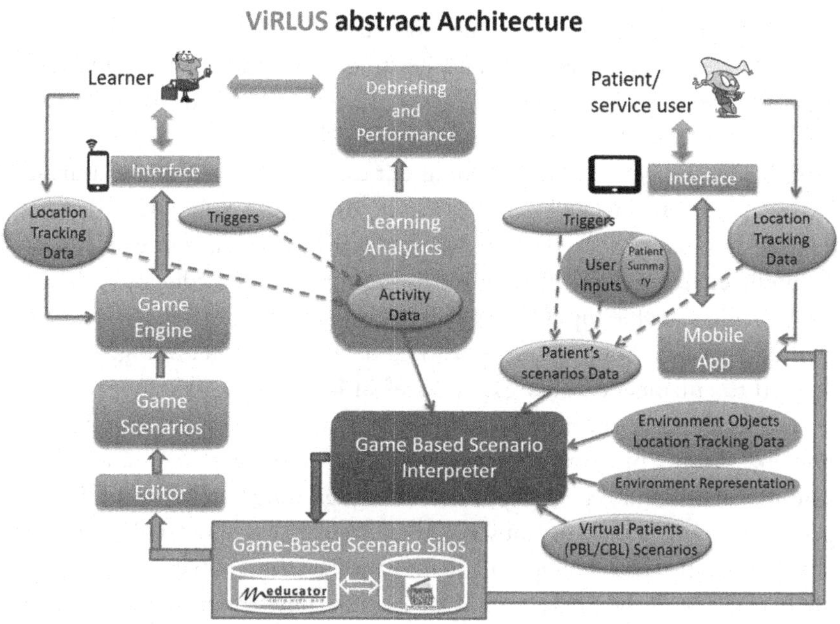

Figure 5.3 *ViRLUS abstract architecture overview.*

- *Reduction.* A classic definition of Reduction in Persuasive technology is "Persuading through simplifying."
- *Tunneling.* The textbook definition of Tunneling is "leading users through a predetermined sequence of actions or events step by step."
- *Tailoring.* A powerful characteristic of persuasive technology is customization, in which "information relevant to individuals to change their attitudes or behaviors" is provided.

Furthermore, a motivational framework developed for online serious games for seniors,[94,95] that can be easily adapted for learners, is utilized by ViRLUS. Following that framework, a number of behavior change triggers are utilized such as:

- Display information to encourage people to be more active.
- Record and display the user's past behavior.
- Use positive reinforcements to improve behaviors.
- Make an attractive and friendly user interface.
- Provide information at opportune moments.
- Use social influence.
- Personalization.

Conclusion

Information and communication technologies are incremental engaged in healthcare education and existing applications and initiatives of pervasive learning,[96,97] context aware learning,[98,99] and location-based game[81,100,101] form the first applications of IoTiE.

IoTiE is expected to impact different aspects of teaching and learning. Experiential learning will become more cost effective and will allow students to link their theoretical knowledge with practice, in a safe "real" environment, enabling efficient resource utilization. Existing education theories can provide valuable input to the design of IoTiE applications, while tested gaming theories, behavioral theories, and persuasive technologies can be utilized to enhance learners' experiences in IoTiE sessions.

IoTiE could be informed also by student activities before starting a session, like visiting placements or hospitals, knowledge acquired online, books lending form library, and others, etc. Using learning analytics and decision making,[31,40] the IoTiE sessions could be personalized and tailored to the needs of each user. Having said that, learners with special needs[102] can be further helped as the "system" will know their physical or mental boundaries and will instantly adjust it to their needs.

IoTiE is also expected to impact industry and infrastructure in relation with future education. Currently the global e-Learning market is expected to reach $325 billion by 2025,[e] while the serious game for education itself has a growth rate over 30%,[f] revealing a quick adoption of digital innovation in healthcare. Furthermore, in order IoTiE to be realized in practice, Higher Education Institutions have to invest on new infrastructure, such as sensors, geolocation systems, transform existing objects to smart Things, etc., as well as, learning analytics and IoT applications.

While the impact of IoT in education will be high, a number of challenges, both at technical level and on theoretical level have to be resolved. As the technology for IoT is in its infancy security issues can easily arise,[49,54] but solutions are being proposed.[50] Furthermore, users' education on cybersecurity including IoT should be enhanced, as it is believed that they are the primary cause of security breaches.[103]

It should be also noted that regulations and laws around the IoT should be enhanced. A lot of countries have already released regulations toward that direction, but further steps needs to be done.[54] The ethical use and approach on IoT created data and applications should be guarded and be ensured through appropriate agreements and technologies.[104,105]

With a range of IoT technologies continuously tested and implemented following interoperability, standards and common architectural frameworks, IoTiE will be soon become available to a wider audience of learning technologist, teachers, and learners. It is believed that IoT will be transformed to IoE, where everything will connected though some kind of sensor communicating with each other without the need of human intervention. IoT or IoEiE is pivotal for a personalized and experiential Healthcare Education.

References

1. Parameshvara Deva M. Use of videotapes in psychological medicine. *Med J Malaysia* 1981;**36**:268–71.
2. Warren M. *Our shared legacy: nursing education at Johns Hopkins, 1889-2006.* Baltimore, MD, USA: Johns Hopkins University Press; 2016.
3. Masic I, Pandza H, Toromanovic S, et al. Information technologies (ITs) in medical education. *Acta Inform Med* 2011;**19**:161–7.
4. Bamidis PD, Konstantinidis S, Papadelis CL, et al. An e-learning platform for aerospace medicine. *Hippokratia* 2008;**12**(Suppl. 1):15–22.
5. Yau JY-K, Joy M, Dickert S. A mobile context-aware framework for managing learning schedules—data analysis from a diary study. *Educ Technol Soc* 2010;**13**:22–32.

[e] https://www.researchandmarkets.com/research/qgq5vf/global_elearning
[f] http://gsvadvisors.com/wordpress/wp-content/uploads/2012/04/GSV-EDU-Factbook-Apr-13-2012.pdf

6. Bamidis PD, Konstantinidis ST, Kaldoudi E, et al. New approaches in teaching medical informatics to medical students. In: *21st, IEEE., international symposium on computer-based medical, systems,* I.E.E.E.; 2008. p. 385–390.

7. Konstantinidis S, Kaldoudi E, Bamidis P. Enabling content sharing in contemporary medical education: a review of technical standards. *J Inf Technol Healthc* 2009;**7**:363–75.

8. Konstantinidis ST, Ioannidis L., Spachos D., et al. mEducator 3.0, combining semantic, social web approaches in sharing, retrieving medical education resources. In: *Proceedings of 2012 seventh international workshop on semantic, social media adaptation, personalization* (SMAP 2012). Luxemburg, Luxemburg; 2012, p. 42–47.

9. Bamidis PD, Konstantinidis ST, Bratsas C, et al. Federating learning management systems for medical education: A persuasive technologies perspective. In: *2011 24th international symposium on computer-based medical systems (CBMS).* IEEE; 2011, p. 1–6.

10. Kaldoudi E, Konstantinidis S, Bamidis P. Web 2.0 approaches for active, collaborative learning in medicine and health. In: Mohammed S, Fiaidhi J, editors. *Health and medical informatics: the ubiquity 2.0 trend and beyond.* Hershey, PA, USA: IGI Global; 2010.

11. Kaldoudi E, Konstantinidis S, Bamidis PD. *Web advances in education: interactive, collaborative learning via web 2.0*; 2010. DOI: 10.4018/978-1-60566-940-3.ch002.

12. Paton C, Bamidis PD, Eysenbach G, et al. Experience in the use of social media in medical and health education, contribution of the IMIA social media working group. *Yearb Med Inform* 2011;**6**:21–9.

13. Cooper JB, Taqueti VR. A brief history of the development of mannequin simulators for clinical education and training. *Qual Saf Health Care* 2004;**13**(Suppl. 1):i11–8.

14. Dafli E, Antoniou P, Ioannidis L, et al. Virtual patients on the semantic Web: a proof-of-application study. *J Med Internet Res* 2015;**17**:e16.

15. Poulton T, Balasubramaniam C. Virtual patients: a year of change. *Med Teach* 2011;**33**:933–7.

16. Kiili K. Digital game-based learning: towards an experiential gaming model. *Internet High Educ* 2005;**8**:13–24.

17. Zualkernan IA. Exploring problem posing as a pedagogical device in the context of wearable, tangible and ubiquitous game-based learning. In: Chova, LG, Torres, IC, Martinez A, editors. *INTED 2011: fifth international technology, education and development conference*; 2011, p. 4999–5008.

18. Chen JH, Wang TH, Chang WC, et al. Developing the historical culture course by using the Ubiquitous Game-Based Learning environment. In: Li F, Zhao J, Shih TK, Lau R, Li Q, McLeod D, editors. *Advances in Web based learning—ICWL 2008, Proceeding.* S; 2008, p. 241–252.

19. Chang Y-H, Lin B-S. A ubiquitous system supporting game-based inquiry learning. *Int J Mob Commun* 2012;**10**:190–212.

20. Chang W-C, Wang T-H, Lin FH, et al. Game-based learning with ubiquitous technologies. *IEEE Internet Comput* 2009;**13**:26–33.

21. Konstantinidis ST, Bamidis PD. E-Learning environments in medical education: how pervasive computing can influence the educational process. In: Puneet S, editor. *Pervasive computing: a multidimensional approach.* India: ICFAI Books; 2009.

22. Miorandi D, Sicari S, De Pellegrini F, et al. Internet of things: vision, applications and research challenges. *Ad Hoc Networks* 2012;**10**:1497–516.

23. Weiser M. The computer for the 21st Century. *Sci Am.* 1991. DOI: https://doi.org/10.1038/scientificamerican0991-94.

24. Satyanarayanan M. Pervasive computing: vision and challenges. *IEEE Pers Commun* 2001;**8**:10–7.

25. Ashton K. That 'Internet of Things' thing. *FRID J* 2009. Available from: http://www.rfidjournal.com/articles/pdf?4986.

26. Gubbi J, Buyya R, Marusic S, et al. Internet of Things (IoT): a vision, architectural elements, and future directions. *Futur Gener Comput Syst* 2013;**29**:1645–60.

27. Minerva R, Biru A, Rotondi D. Towards a definition of the Internet of Things (IoT); 2015. Available from: http://iot.ieee.org/images/files/pdf/IEEE_IoT_Towards_Definition_Internet_of_Things_Revision1_27MAY15.pdf.

28. Atzori L, Iera A, Morabito G. The Internet of Things: A survey. *Comput Networks* 2010;**54**:2787–805.

29. Behmann F, Wu K. *Collaborative internet of things (C-IoT): for future smart connected life and business.* Chichester, England: IEEE: Wiley; 2015.

30. Selinger M, Sepulveda A, Buchan J. *Education and the Internet of Everything*; 2013. Available from: https://www.cisco.com/c/dam/en_us/solutions/industries/docs/education/education_internet.pdf.

31. Konstantinidis ST, Bamidis PD. Why decision support systems are important for medical education. *Heal Technol Lett* 2016;**3**:56–60.

32. Kulkarni A, Sathe S. Healthcare applications of the Internet of Things: a review. Available from: http://citeseerx.ist.psu.edu/viewdoc/download?doi=10.1.1.659.5696&rep=rep1&type=pdf. (accessed 12 February 2018).

33. Simonov M, Zich R, Mazzitelli F. *Personalized healthcare communication in Internet of Things*; 2008.

34. Ahmadi H, Arji G, Shahmoradi L, et al. The application of internet of things in healthcare: a systematic literature review and classification. *Univers Access Inf Soc* 2019;**18**: 837–69.

35. Riazul Islam SM, Daehan Kwak, Humaun Kabir M, et al. The Internet of Things for Health Care: a comprehensive survey. *IEEE Access* 2015;**3**:678–708.

36. Konstantinidis EI, Bamparopoulos G, Billis A, et al. Internet of Things for an age-friendly healthcare. *Stud Health Technol Inform* 2015;**210**:587–91.

37. Konstantinidis EI, Billis AS, Dupre R, et al. IoT of active and healthy ageing: cases from indoor location analytics in the wild. *Health Technol (Berl)* 2017;**7**:41–9.

38. Konstantinidis ST, Billis A, Wharrad H, et al. *Internet of Things in health trends through bibliometrics and text mining*; 2017. DOI: 10.3233 / 978-1-61499-753-5-73.

39. Yin Y, Zeng Y, Chen X, et al. The internet of things in healthcare: an overview. *J Ind Inf Integr* 2016;**1**:3–13.

40. Dimitrov DV. Medical Internet of Things and big data in Healthcare. *Healthc Inform Res* 2016;**22**:156–63.

41. Al-Fuqaha A, Guizani M, Mohammadi M, et al. Internet of Things: a survey on enabling technologies, protocols, and applications. *IEEE Commun Surv Tutorials* 2015;**17**:2347–76.

42. Khan R, Khan SU, Zaheer R, et al. Future Internet: The Internet of Things architecture, possible applications and key challenges. In: *2012 tenth international conference on frontiers of information technology*; 2012, p. 257–260.

43. Guillemin P, Berens F, Vermesan O, et al. *Internet of Things, position paper on standardization for IoT technologies.* Available from: http://www.internet-of-things-research.eu/pdf/IERC_Position_Paper_IoT_Standardization_Final.pdf. 2015 [accessed 22 June 2018].

44. Trappey AJC, Trappey CV, Hareesh Govindarajan U, et al. A review of essential standards and patent landscapes for the Internet of Things: a key enabler for Industry 4.0. *Adv Eng Informatics* 2017;**33**:208–29.

45. Heer T, Garcia-Morchon O, Hummen R, et al. Security challenges in the IP-based Internet of Things. In: *Wireless Personal Communications*; 2011. DOI: 10.1007 / s11277-011-0385-5.

46. O'Neill M. Insecurity by design: today's IoT device security problem; 2016. DOI: 10.1016 / J.ENG. 2016.01.014.

47. Mendez D, Papapanagiotou I, Yang B. Internet of Things: Survey on Security and Privacy. *IOT Secur.*

48. Mahmoud R, Yousuf T, Aloul F, et al. Internet of things (IoT) security: current status, challenges and prospective measures. In: *2015 tenth international conference for internet technology and secured transactions (ICITST)*. London, UK: IEEE; 2015, p. 336–341.

49. Suo H, Wan J, Zou C, et al. Security in the internet of things: a review. In: *Proceedings of 2012 international conference on computer science and electronics engineering, ICCSEE 2012*; 2012. DOI: 10.1109 / ICCSEE. 2012.373.
50. Sicari S, Rizzardi A, Grieco LA, et al. Security, privacy and trust in Internet of Things: the road ahead. *Comput Networks* 2015;**76**:146–64.
51. Lin H, Bergmann N. IoT privacy and security challenges for smart home environments. *Information* 2016;7(3):44. doi: 10.3390 / info7030044.
52. Jing Q, Vasilakos A V., Wan J, et al. Security of the Internet of Things: perspectives and challenges. *Wirel Networks* 2014;20:2481–2501. doi: 10.1007 / s11276-014-0761-7.
53. Vani G, Malakreddy BA. Security challenges in IOT applications in health care domain. *Int J Adv Electron Comput Sci* 2016;(Special Issue):141–4.
54. Weber RH. Internet of things: privacy issues revisited. *Comput Law Secur Rev* 2015;**31**:618–27.
55. Patton M, Gross E, Chinn R, et al. Uninvited connections: a study of vulnerable devices on the internet of things (IoT). In: *Proceedings of 2014 IEEE joint intelligence and security informatics conference, JISIC 2014*; 2014; p. 232–5. doi: 10.1109/JISIC. 2014.43.
56. Durumeric Z, Adrian D, Mirian A, et al. A search engine backed by Internet-wide scanning. In: *Proceedings of the 22nd ACM SIGSAC conference on computer and communications security—CCS' 15*; 2015; p. 542–533. doi: 10.1145 / 2810103.2813703.
57. Lave J. *Jean Lave: cognition in practice: mind, mathematics and culture in everyday life.* Cambridge, UK: Cambridge University Press; 1988.
58. Wenger E. Communities of practice and social learning systems: the career of a concept. In: *Social learning systems and communities of practice*; 2010. p. 179–198. doi: 10.1007/978-1-84996-133-2_11.
59. Lave J, Wenger E. *Situated learning: legitimate peripheral participation.* Cambridge, UK: Cambridge University Press; 1991 DOI: 10.2307 / 2804509.
60. Bandura A. *Social learning theory.* New York: General Learning Press; 1977.
61. Vygotsky L. *Mind in society: the development of higher psychological processes.* Cambridge: Harvard University Press; 1978.
62. Dewey J. *Experience & education. Kappa Delta Pi lecture series.* New York: Macmillan; 1938.
63 Kolb DA. . *Experiential learning: experience as the source of learning and development.* Upper Saddle River, NJ: Prentice Hall PTR; 1984doi: 10.1016/B978-0-7506-7223-8.50017-4.1984.
64. Spiro RJ, Jehng J-C. Cognitive flexibility and hypertext: theory and technology for the nonlinear and multidimensional traversal of complex subject matter. In: Nix D, Spiro R, editors. *Cognition, education, and multimedia: exploring ideas in high technology.* Hillsdale, NJ, US: Lawrence Erlbaum Associates, Inc.; 1990. p. 163–205.
65. Schmidt HG. Foundations of problem-based learing: some explanatory notes. *Med Educ* 1993;**27**:422–3.
66. Gewurtz RE, Coman L, Dhillon S, et al. Problem-based learning and theories of teaching and learning in health professional education. *J Perspect Appl Acad Pract* 2016;**4**:59–70.
67. Arafeh JMR, Hansen SS, Nichols A. Debriefing in simulated-based learning: facilitating a reflective discussion. *J Perinat Neonatal Nurs* 2010;**24**:302–9.
68. Dreifuerst KT. The essentials of debriefing in simulation learning: a concept analysis. *Nurs Educ Perspect* 2009;**30**:109–14.
69. Peterson M. Skills to enhance problem-based learning, medical education online. *Med Educ Online* 1997;**2**:4289.
70. Rudolph JW, Simon R, Raemer DB, et al. Debriefing as formative assessment: closing performance gaps in medical education. *Acad Emerg Med* 2008;**15**:1010–6.
71. Lederman LC. Debriefing: toward a systematic assessment of theory and practice. *Simul Gaming* 1992;**23**:145–60.
72. Koshy K, Limb C, Gundogan B, et al. Reflective practice in health care and how to reflect effectively. *Int J surgery Oncol* 2017;**2**:e20.

73. Fogg BJ. *Persuasive technology: using computers to change what we think and do.* Morgan Kaufmann, 2003. DOI: 10.4017 / gt.2006.05.01.009.00.
74. Chatterjee S, Price A. Healthy living with persuasive technologies: framework, issues, and challenges. *J Am Med Informatics Assoc* 2009;**16**:171–8.
75. King P, Tester J. The landscape of persuasive technologies. *Commun ACM* 1999;**42**:31–8.
76. Kort YAW De, IJsselsteijn WA, Eggen JH, et al. Persuasive gerontechnology. *Gerontechnology* 2005;**4**:123–7.
77. Fogg BJ. Creating persuasive technologies: an eight-step design process. *Technology* 2009;**91**:1–6.
78. Oinas-Kukkonen H, Harjumaa M. Persuasive systems design: key issues, process model, and system features. *Commun Assoc Inf Syst* 2009;**24**:485–500.
79. Hawkins RP, Kreuter M, Resnicow K, et al. Understanding tailoring in communicating about health. *Health Educ Res* 2008;**23**:454–66.
80. Klopfer E, Squire K. Environmental detectives—the development of an augmented reality platform for environmental simulations. *Educ Technol Res Dev* 2008;**56**:203–28.
81. Slussareff M, Simkova Z. Location-based games enhancing education: design and implementation lessons learnt. In: Research AI for E and (ed). Athens: Athens: ATINER'S Conference Paper Series, 2014, p. No: EDU2014-0980.
82. Avouris N, Yiannoutsou N. A review of mobile location-based games for learning across physical and virtual spaces. *J Univers Comput Sci* 2012;**18**:2120–42.
83. Gustafsson A, Bichard J, Brunnberg L, et al. Believable environments: generating interactive storytelling in vast location-based pervasive games. In: *Proceedings of the 2006, ACM., SIGCHI., international conference on advances in computer entertainment technology— ACE., '06.*, New York, New York, USA: ACM Press, p. 24.
84. Wijers M, Jonker V, Drijvers P. MobileMath: exploring mathematics outside the classroom. *ZDM—Int J Math Educ* 2010;**42**:789–99.
85. Colella V. Participatory simulations: building collaborative understanding through immersive dynamic modeling. *J Learn Sci* 2000;**9**:471–500.
86. Walker K. Visitor-constructed personalized learning trails. In: Trant J, Bearman D, editors, *Proceedings of Museums and the Web* 2007. Toronto: Archives & Museum Informatics, 2007.
87. Akkerman S, Admiraal W, Huizenga J. Storification in history education: a mobile game in and about medieval Amsterdam. *Comput Educ* 2009;**52**:449–59.
88. Benford S, Magerkurth C, Ljungstrand P. Bridging the physical and digital in pervasive gaming. *Commun ACM* 2005;**48**:54–7.
89. Facer K, Joiner R, Stanton D, et al. Savannah: mobile gaming and learning? *J Comput Assist Learn* 2004;**20**:399–409.
90. Boulos MNK, Yang SP. Exergames for health and fitness: the roles of GPS and geosocial apps. *Int J Health Geogr* 2013;**12**:18.
91. Ahmed M, Sherwani Y, Al-Jibury O, et al. Gamification in medical education. *Med Educ Online* 2015;**20**:29536.
92. Gorbanev I, Agudelo-Londoño S, González RA, et al. A systematic review of serious games in medical education: quality of evidence and pedagogical strategy. *Med Educ Online* 2018;**23**:1438718.
93. Kopeć W, Abramczuk K, Balcerzak B, et al. A location-based game for two generations: teaching mobile technology to the elderly with the support of young volunteers. In: Giokas K, Bokor L, Hopfgartner F, editors, eHealth 360°. *Lecture notes of the Institute for Computer Sciences, Social Informatics and Telecommunications Engineering*, vol 181. Springer, Cham, p. 84–91.
94. Konstantinidis ST, Brox E, Kummervold PE, et al. *Online social exergames for seniors: a pillar of gamification for clinical practice*; 2015. DOI: 10.4018 / 978-1-4666-9522-1.ch012.
95. Bamidis PD, Gabarron E, Hors-Fraile S, et al. *Gamification and behavioral change: techniques for health social media*; 2016. DOI: 10.1016 / B978-0-12-809269-9.00007-4.

96. Bicocchi N, Mamei M, Prati A, et al. Pervasive self-learning with multi-modal distributed sensors. In: Serugendo, GD, editor, *SASOW 2008: second IEEE international conference on self-adaptive and self-organizing systems workshops, proceedings*; 2008, p. 61–66.
97. Suo Y, Shi Y. Towards blended learning environment based on pervasive computing technologies. In: Fong J, Kwan R, Wang FL, editors, *Hybrid learning and education, proceedings*; 2008, p. 190–201.
98. Tortorella RAW. Kinshuk. A mobile context-aware medical training system for the reduction of pathogen transmission. *Smart Learn Environ* 2017;**4**:4.
99. Wang S-L, Chen C-C, Zhang ZG. A context-aware knowledge map to support ubiquitous learning activities for a u-Botanical museum. *Australas J Educ Technol* 2015;**31**:470–85.
100. Kasaki N, Kurabayashi S, Kiyoki Y. A geo-location context-aware mobile learning system with adaptive correlation computing methods. In: Shakshuki E, Younas M, editors, *ANT 2012 and MOBIWIS 2012;* 2012, p. 593–600.
101. Towards location-based augmented reality games. *Procedia Comput Sci* 2012; 15: 318–319.
102. McRae L, Ellis K, Kent M. *Internet of Things (IoT): Education and Technology; The relationship between education and technology for students with disabilities*, https://www.ncsehe.edu.au/wp-content/uploads/2018/02/IoTEducation_Formatted_Accessible.pdf (2018).
103. Choi M, Levy Y, Anat H. The role of user computer self-efficacy, cybersecurity countermeasures awareness, and cybersecurity skills influence on computer misuse. In: *Proceedings of the Pre-International Conference of Information Systems (ICIS) SIGSEC - Workshop on Information Security and Privacy (WISP) 2013*. Milan, https://nsuworks.nova.edu/gscis_facpres/98 (2013, accessed 26 June 2018).
104. Baldini G, Botterman M, Neisse R, et al. Ethical Design in the Internet of Things. *Sci Eng Ethics* 2018;**24**:905–25.
105. Lally V, Sharples M, Tracy F, et al. Researching the ethical dimensions of mobile, ubiquitous and immersive technology enhanced learning (MUITEL): a thematic review and dialogue. *Interact Learn Environ* 2012;**20**:217–38.

CHAPTER SIX

Social web and social media's role in healthcare education and training

Neil Withnell
University of Salford, Manchester, England

Chapter outline

Introduction

Think about the information that you gathered today. How much of it came from social media such as Twitter, YouTube, Facebook, or any of the many other social media sites? Consider also what you then did with the information that you had. Who you have shared it with and to what end? The likelihood is that for many people, a lot of the information they garnered and shared involved the use of social media. This ranges from personal information, sharing thoughts and ideas, to learning about the news, and what is happening in our localities and around the world. Harnessing this connectivity, and the vast audience that social media gives access to, is increasingly important in both health care provision and healthcare education. So what is social media and what are its implications for health care education and training? Social media may already be a large part of many people's lives but how are they using this? With many healthcare organizations using social media, the question should now be how to make the best use of social media.

Digital Innovations in Healthcare Education and Training.
http://dx.doi.org/10.1016/B978-0-12-813144-2.00006-4
Copyright © 2021 Elsevier Inc. All rights reserved.

Background

Takahashi et al.[1] referred to a revolution in the field of communication through the internet, referred to as Web 2.0. Takahashi and colleagues describe Web 2.0, such as Facebook and other Social Network Services (SNSs) as applications that allow the building of online social networks where individuals can share interests and activities. They also refer to SNSs for specific heath related purposes such as quitting smoking. There are increasing SNSs in healthcare and it is inevitable that healthcare education follows suit. Healthcare educators are already using social media and more research will follow as people seek to gauge the efficacy of social media use. As with all innovations, there are benefits and potential pitfalls and some of these will only become apparent as research into this area develops.

Lee Ventola[2] looked at the benefits, risks, and best practices in relation to social media for healthcare professionals. The work undertaken describes various social media sites:

- Social networking (Facebook, Google Plus)
- Professional networking (LinkedIn)
- Media sharing (YouTube)
- Content production blogs(Blogger Tumblr)
- Microblogs (Twitter)—the word blog deriving from web log
- Knowledge/information (Wikipedia)
- Virtual Reality and gaming environments (Second Life)

Lee Ventola[2] states that in the United States, the proportion of adults using social media has increased from 8% to 72% since 2005 and that one seventh of the world's population are on Facebook. In the United Kingdom, it is estimated that in 2016, 78% of all women and 73% of men had a social media account (Statista[3]). The forecast from this data is that in the United Kingdom by 2022, the number of monthly active social network users is projected to reach 42.88 million individuals, which is an increase of over 4 million new users from 38.01 million in 2015. Lee Ventola[2] also makes reference to the fact that social media plays a part in significant political events and uses the Arab Spring as evidence of this. Presently there are negotiations occurring between the UK home secretary and the internet giants to look at ways of managing national security in how to view terrorism and the use of social media by terrorists to communicate. This topic is hotly debated as many people see measures to address this as potential limitations on freedom of speech and we will come back to this point later. So on a wide scale, the use of social media can be seen as something with benefits and disadvantages. So it follows that this will be the case when trying to utilise social media in healthcare education.

A comprehensive study by Price Waterhouse Cooper, Health Research Institute (HRI)[4] identified that health was slower to move on the potential presented by social media but they have followed in the footsteps of retail and hospitality. At present there are more than 9 out of 10 NHS organizations who use some form of social media. Prime examples of health organizations using Twitter, for example, include the Department of Health & Social Care (@DHSCgovuk) with over 665,000 followers. NHS England (@NHSEngland) have over 424,000 followers and the King's Fund (an independent charity, @TheKingsFund) with over 144,000 followers. These are significant Twitter accounts with a wide reach, and undoubtedly have an impact in healthcare, check out your local hospital and see how effectively they are using social media.

The study further comments that "With transparency patient expectations rise." If this is the case in healthcare then can we assume that with transparency in healthcare education and the use of social media students expectations will rise and will this be a driver for enhanced teaching? It would not be unreasonable for avid users of social media to look at education programmes and use of social media before committing to learning with that institution.

They identified age as the most influential factor in engaging and sharing through social media with more than 80% of individuals aged 18–24 being likely to share health information through social media and nearly 90% engaging in health activities or trusting information found via social media. Less than half (45%) of individuals aged 45–64 would be likely to share via social media and 56% of this group would be likely to engage in health activities. Educators need to attract learners to be viable businesses and if you target audience expect social media as part of their learning then educators need to be able to demonstrate how they incorporate this into their curriculum. More importantly they need to be measuring how effective social media is in enhancing learning for students.

They also describe social media as an instantaneous communication channel which changes online dialogue from one-to- many to many-to-many at a phenomenal speed.

HRI[4] breaks social media into four components; user generated content; community; rapid distribution; and open two-way dialog. Facebook is a perfect example of this, and Facebook is renowned for being the number one social media platform.

With reference to the effects of social media in health, HRI[4] states "Social networks will peel back every corner of the health system and drive transparency on cost, value and outcomes." If this is the case then the impact upon education may also lead to more transparency which can potentially drive up standards, and more importantly give learners more of a voice.

Don Sinko, Chief Integrity Officer at Cleveland Clinic is quoted by HRI[4]; "One of the greatest risks of social media is ignoring social media. It's out there and people are using it whether you like it or not. You don't know what you don't know. Listening is the start to knowing." The study also presents an example that demonstrates this; a network of nurses on a social media site were discussing defects of a certain drug. The drug maker's executives had no knowledge of the defects or the nurse's discussions until the "chatter" was discovered by regulatory authorities on one of the drug company's social media sites. After this experience, the company quickly established a capability to mine information from the social online community. Education could more effectively "mine" for understanding of the student experience via social media with a view to keep the students at the center of planning education.

HRI[4] suggests that "social media has the ability to pull together a fragmented (healthcare) industry, with the patients and their information in the centre."

HRI[4] presents a social media participation model for businesses which has three steps:

- Listen—actively monitor and capture conversation to analyze and understand the meaning of what is being said, the sentiment of the discussion, and what influence it has over audiences.
- Participate—proactively post and publish content on social media enabled platforms to communicate a message to an audience, but not necessarily engage them in conversation.
- Engage—actively interact in one-to-one, one-to-many, or many-to-many conversations within social media in order to freely exchange information and advance discussion.

This business model is useful for all organizations that want to harness the power of social media. As healthcare and education in healthcare, increasingly move toward outcomes based reimbursement, social media can provide essential links and platforms.

Kietzmann et al.[5] identified the changing role of the internet. He postulated that people had developed beyond using the internet for reading, watching, and buying into usage that focused on creating, modifying, and

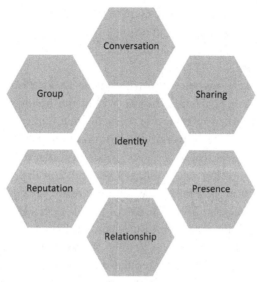

Figure 6.1 *Functional building blocks.*

sharing. He developed a framework to define social media using seven functional building blocks (Fig. 6.1).

Each block examines a specific of social media user experience.

Identity represents the extent to which users reveal their identities. This can include information such as name, age, gender, profession, and location. The extent to which users want to reveal personal information will vary. Many users have user names as oppose to their real names. In terms of health related information and healthcare education, users may want to restrict personal information if service users are also accessing the social media. This would help to maintain professional boundaries.

Conversation represents the extent to which users communicate with other users. Is the conversation intended to be short and punchy or is it a forum for users to make long detailed postings? If organisations can manipulate the conversations to suit their marketing, as Kietzmann et al.[5] describes, then can educators set the scene for the types of conversation they feel will be most conducive to learning?

Sharing represents the extent to which users exchange, distribute, and receive content. This looks at what users have in common and what they potentially will have in common and want to share. This "block" also questions the degree to which the object can or should be shared. Kietzmann and colleagues[5] gives the example of YouTube which faced lawsuits for failing to ensure that uploaded content complied with copyright laws. Within an education context, attention needs to be paid to sharing. The information should be accurate and there needs to be oversight of who it can be shared with.

Presence refers to the extent to which users know if other users are accessible. Kietzmann et al.[5] links blocks within the honeycomb model and presence is linked to conversation in the belief that higher levels of presence are more likely to make conversations more influential.

Relationship refers to the forms of association that then lead people to converse and have online relationships. These relationships can be loose and informal or much more regulated. Within the realm of education there will be the need for regulation as the use of social media will be for specific outcomes that will need to be measured and evaluated.

Reputation refers not only to people but to their content. An education establishment would want a reputation of aiding learning. Number of followers or users can indicate reputation as can likes and online endorsements. Users want to protect their own online reputations and will not use a site that jeopardizes this.

Group relates to the extent to which users can form communities and sub-communities. Also who communicates with who? Within groups, there may be different sharing permissions for users, some users may be having greater access to the whole group and some having restrictions. Within an education context more advanced learners may have increased access to others whereas novice learners may have some restrictions until their knowledge base is increased.

The blocks are useful for anybody trying to understand social media. The purpose of Kietzmann[5] and colleagues' work was to help organizations develop strategies for monitoring, understanding, and responding to different social media activities. The framework was intended to help organizations feel more comfortable with the use of social media to improve business. It had been identified that some executives were eschewing social media to their organization's detriment. Healthcare education and training could also benefit from the work as there are those that embrace the use of social media and those who shy away from it. The latter are potentially decreasing as learners from all walks of life seek to extend their learning via social media channels and educators need to keep up.

Cheston et al.[6] comments on the fast paced evolvement of new technologies and acknowledges that while this is a challenge "the pace can also offer opportunities for innovation, such as engaging learners in curricular activity." Ignoring social media in education simply does not seem to be an option so better to embrace it with an awareness of potential pitfalls as the technologies develop further.

Use of social media in healthcare education and training

The important question is how do we harness the use of social media in healthcare education?

Moorhead et al.[7] looked at the benefits and limitations of social media for health communication and these are pertinent to the use of social media in healthcare education. Benefits cited included the sharing of tailored information, increased interactions, and widened access. Blogs were viewed positively as individuals could access tailored resources to deal with health issues. In a similar way, blogs are useful in education as a student can identify their own interests and weaknesses and seek tailored responses. Gunther's work identified that "an important aspect of using social media for health communication is that it can provide valuable peer, social, and emotional support....," all of which transfer neatly to healthcare education.

The limitations found by Moorhead and colleagues[7] were most notably quality concerns and the lack of reliability. As previously mentioned when referring to Kietzmann,[5] this indicates a need for an element of control over the process within an educational context.

Kietzmann et al.[5] indicated that social media is changing the relationship between producers and consumers of a message, and as such suggested that healthcare providers may need to take a certain degree of control over online health communication to maintain validity and reliability. This is also key in healthcare education as unreliable information could be quickly and extensively shared.

Blogs are useful in education and are the oldest form of social media. They reach a large audience (especially if posts go viral) and they allow readers to leave comments which enables dialogue which can then be conducive to further learning. Lee Ventola[2] reports that nearly two thirds of blogs are written anonymously. This can present difficulties as there is no accountability for the veracity of what is said. Conversely, it can allow people to ask questions and try out theories without fear of being judged as ignorant.

The far reaching nature of sites such as Twitter cannot be underestimated. Fogelson et al.[8] cites Twitter as a tool that has potentially affected health policy decisions with physicians developing large followings and reaching broad audiences that policy makers want to engage with. Twitter is instant and a "conversation" can happen very quickly on this platform, with many followers able to share the news. During a public health crisis twitter (along with other social media platforms) can quickly spread the word, both informing and protecting the public.

Wikipedia has been referred to as the most commonly used wiki in the medical community. It can be somewhat disconcerting when a General Practitioner "googles" your symptoms and quotes Wikipedia to you, especially as you may well have consulted Wikipedia before going to the Doctors surgery. Google searches heavily promote Wikipedia and it can be a useful source of information. Not everybody realizes that the information is not verified and that it can be edited and added to. This can result in information being updated and made more accurate but it also leaves a huge window for incorrect information to be further disseminated. We need to look at sites such as Wikipedia in terms of education as students will refer to it. Is it a short cut to information? More importantly is it a short cut to incorrect information? These issues need to be addressed in helping learners develop robust research skills.

Lee ventola[2] talks about social media improving clinical education by way of its inherent communication capabilities and cites studies of social media "which have enhanced students understanding of communication, professionalism, and ethics." An interesting reference in the Lee Ventola[2] study is to Auburn University where instructors established Twitter handles so that pharmacy students could participate in class discussion anonymously. While 81% of students felt that it helped them express opinion they otherwise would not have, 71% thought Twitter had been distracting. The experience of social media can be seen as individual in this context, but this also open up the potential for the virtual classroom, with many people being able to engage in discussions (such as the popular Twitter chats).

The question of anonymity is interesting. Does it allow more confident exploration of issues without fear of being seen as ignorant or does it encourage less robust self-reflection before opinions are put into the public domain?

Lee Ventola[2] talks about balancing increased communication opportunities against increased distraction in an educational environment and the challenge this presents. Lee Ventola[2] also makes the point that the standards guiding appropriate use of social media are in their infancy. This is discussed later in this chapter.

Role of educators

Educators have a responsibility to make students aware of the limitations of social media to avoid inaccurate information being disseminated. Other limitations involve privacy and confidentiality. The use of social media is essentially a collaborative process and as such all parties have a responsibility for the information and how it is used. If learners are encouraged to reflect and potentially disclose feelings and thoughts they should be aware of who this information is open to and whether they want such exposure.

Whether social media is embraced by more people will depend upon dealing with the limitations. One obstacle to wider use within several fields is the differing abilities of people with new technologies. Consideration needs to be given to less confident people (learners and educators alike) and training made accessible to increase confidence in using social media.

Social media can be empowering to learners because of its collaborative nature. Engaging with people who can respond with immediacy helps learners to set their own agendas in terms of what they want from that interaction and how they intend to achieve their desired outcome. Social media provides communication in real time so learners and educators can be engaged in a dynamic process that is not owned by one person. There are also cost implications in this as they are vastly reduced with people remaining in their own localities while communicating with people around the country and the world. However social media is unregulated and therefore educators have a responsibility to learners to look at controls and safety mechanisms. Learners also need to be forewarned about potential pitfalls. Some nursing careers have been ended before they have scarcely begun because of inappropriate comments on social media, examples of students thinking they are being reflective when they are actually breaching confidentiality.

Professionalism within the disciplines

The regulatory body for nursing and midwifery in the United Kingdom, The (NMC) Nursing and Midwifery Council[9] has guidelines of using social media which runs alongside the Code of Conduct. The principles are succinct and useful for all users of social media within health care education. The guidelines advise that nurses can put their registration at risk and students can impact on their ability to join the register if they act in a way that is unprofessional or unlawful on social media. The guidance advises:

- sharing confidential information inappropriately
- posting pictures of patients and people receiving care without their consent
- posting inappropriate comments about patients
- bullying intimidating or exploiting people
- building or pursuing relationships with patients or service users
- stealing personal information or using someone else's identity
- encouraging violence or self-harm
- inciting hatred or discrimination

The NMC advises to be informed and to familiarize yourself with how individual social media applications work, being clear about the advantages and disadvantages. The guidance alerts you to think before you post and realize that even the strictest privacy settings have imitations and once something is online it can be copied and redistributed. It further advises that you protect your professionalism and reputation. As with all things in life it is wise to choose your bedfellows well and the internet is far from the exception.

The (HCPC) Health and Care Professions Council[10] regulates 16 professions including social workers, occupational therapists, dieticians, paramedics, and many other health related professionals. The HCPC produce standards for their registrants in relation to the use of social media and also top tips for the use of social media which are outlined here:

- apply the same standards of behavior as you would elsewhere. If you would not put something in a letter or email or say it out loud, don't say it on social media.
- think before you post. Try to be polite and steer clear of inappropriate or offensive language.
- think about who can see you share and consider managing your privacy settings accordingly.
- maintain appropriate professional boundaries if you communicate with service users.
- maintain appropriate professional boundaries if you communicate with service users or carers.
- do not post confidential or identifiable information.
- do not post inappropriate or offensive material.
- if you are employed, be aware of your employer's social media policy.
- when in doubt, seek advice from a friend or colleague. You can contact HCPC if you are uncertain about the standards.
- keep on posting! Social media is a great communication tool. There's no reason why registrants shouldn't keep on using it with confidence.

The final tip on that list is encouraging as often guidance can be so risk averse that it can make people fearful of the subject matter. For many of us, the use of social media is well established in our day to day lives and it would present a missed opportunity not to exploit this enormous connectivity in our workplaces. Registrants have told the HCPC that social media helps them to develop and share their skills and knowledge, it allows a greater profile of their profession.

However, what you say, or like, or follow on social media can lead to perceptions of you that may damage your professional reputation and this should be kept in mind.

Privacy on social media is an important matter but as a health professional, you may not be able to rely on concepts such as freedom of speech if something you post lies outside the professional boundaries of your regulating body (Peck[11]). Freedom of speech was used as a defence in 2009 in the United States of America. A nursing student was expelled from the course due to posting remarks about the race, sex, and religion of patients under her care. The nursing school expelled her but she put forward a defence that her freedom of speech had been violated. The court rejected this because the school had an "honor code and confidentiality agreement" which she had clearly violated.

While encryption exists on some social media (and this is contentious in terms of current governments trying to work with internet giants to combat the use of social media in terrorism), a professional would be wise to view any postings as potentially in the public domain.

One aspect of social media that can be overlooked but which is of real importance is safety. Dowdell et al.[12] looked at online social networking patterns among adolescents, young adults, and sexual offenders and examined potentially unsafe encounters between predators and the vulnerable or young. The research found that more than half of internet offenders disguised their identities when online. This emphasizes the need for filters and privacy options as whatever the intended audience it could be infiltrated by people with their own agenda. Dowdell[12] suggests "possible nurse initiated policy recommendations include designing technologies and educational programs to help in the identification of suspicious online behaviors." Adult learners may not be seen as vulnerable but we all have vulnerabilities depending upon what others want from us. Disguised identities could facilitate inadvertent breaches of confidentiality or plagiarism.

An interesting aspect of social media is the democratizing aspect. Kietzmann et al.[5] commented that with the rise in social media, it appeared that corporate communication had been democratized and the power had been taken from those in marketing and public relations by the individuals and communities that created, shared, and consumed blogs. This shift in power is interesting in education terms. It gives learners more of a voice and evaluation of learning can help to shape curriculums as students going through a process comment on the efficacy of the methods used in their learning environments. Social media also gives learners the opportunities to comment on what works for them and what does not. This constructive criticism could have the potential to shape future learning.

Kietzmann and colleagues[5] said firms would have to decide if they wanted to get serious about social media or continue to ignore it. He felt both options could have tremendous impact. It is fair to say that most successful businesses now use and benefit from social media. It is likely that the future of education will similarly embrace it and benefit, or, ignore it and suffer.

Conclusion

There is a growing trend to use social media and this has huge potential in healthcare and in healthcare education. Organizations are starting to see the benefits of engaging in this media and while everything that is put online cannot be wholly trusted, with a degree of caution there is undoubtedly a part to play. The way to reach vast audiences is there and surely the whole purpose now is not to consider why to use social media but how to use social media.

References

1. Takahashi Y, Uchida C, Miyaki K, Sakai M, Shimbo T, Nakayama T. Potential benefits and harms of a Peer Support Social Network Service on the Internet for people with Depressive Tendencies: qualitative content analysis and social network analysis. *J Med Internet Res* 2009;**11**(3):e29.
2. Lee Ventola C. Social media and health care professionals: benefits, risks, and best practices. *Pharm Therapeut* 2014;**39**(7):491–9.
3. Statista; the statistics portal. *Forecast of social network user numbers in the United Kingdom (UK) from 2015 to 2022 (in million users)*. Available from: https://www.statista.com/statistics/553530/predicted-number-of-social-network-users-in-the-united-kingdom-uk/; 2017 [accessed at 27th November 2017].
4. Price Waterhouse Cooper: Health research Institute. *Social media "likes" healthcare: From marketing to social business*. Available from: https://www.pwc.com/us/en/health-industries/health-research-institute/publications/health-care-social-media.html; 2012 [accessed 27th November 2017].
5. Kietzmann JH, Hermkens K, McCarthy IP, Silvestre BS. Social Media? Get serious! Understanding the functional building blocks of social media. *Bus Horiz* 2011;**54**:241–51.
6. Cheston CC, Flickinger TE, Chisolm MS. Social media use in medical education: a systematic review. *Acad Med* 2013;**88**(6):893–901.

7. Moorhead SA, Hazlett DE, Harrison L, Carroll JK, Irwin A, Hoving C. A new dimension of healthcare: systematic review of the uses, benefits, and limitations of social media for health communication. *J Med Internet Res* 2013;**15**(4):e85.
8. Fogelson NS, Rubin ZA, Ault KA. Beyond likes and tweets: an in-depth look at the physician social media landscape. *Clin Obstetr Gynecol* 2013;**56**(3):495–508.
9. Nursing and Midwifery Council. *Social media guidance.* Available from: https://www.nmc. org.uk/standards/guidance/social-media-guidance/; 2017 [accessed 27th November 2017].
10. Health and Care Professions Council. *Use of social networking sites.* Available from: http://www.hpc-uk.org/registrants/standards/socialnetworking/; 2017 [accessed 27th November 2017].
11. Peck JL. Social media in nursing education: responsible integration for meaningful use. *J Nurse Edu* 2014;**14**(19):1–6.
12. Dowdell EB, Burgess AW, Flores JR. Original research: online social networking patterns among adolescents, young adults, and sexual offenders. *Am J Nursing* 2011;**111**(7):28–36.

PART THREE

Implementing Digital Innovations

CHAPTER SEVEN

Implementing digital learning for health

Panagiotis E. Antoniou
Medical Physics Laboratory, Medical School, Aristotle University of Thessaloniki, Thessaloniki, Greece

Chapter outline

Core precepts I: medical education discourse and the molding of healthcare experts

While it is tempting to initiate a discourse on the rationale of technology enhancement of learning with a historical time-lapse of technology's incorporation into education, this chapter shall attempt a different approach. In our presentation of the reasons that lead one to utilize digital means for healthcare we shall first tap into works of medical education in order to illustrate how technology organically evolved into healthcare learning instead of being shoehorned into it.

In that attempt Flexner's recommendations,[1] even though dating back to the start of the previous century, stands strong, at least in principle. These form a pedagogical foundation that both considers the status quo of

Digital Innovations in Healthcare Education and Training.
http://dx.doi.org/10.1016/B978-0-12-813144-2.00007-6

medical education, but also heeds the core precepts that have defined (even ad-hoc at times) this practical status quo. Integrative medical education that stemmed from two pillars (pre-clinical theoretical learning and clinical skills acquisition training) defined the axes that would apply almost universally to medical students. First the prospecting doctors would need to be taught the science that was the foundation of their vocation and then they should, in equal measure, be taught the craft of their vocation, that is, the acquisition of clinical skills that would make them competent implementers and decision makers, in short medical experts, for diagnosing and treating their patients. In that context, it was expected that knowledge should be transferred prior to application of its effects and that learning the craft and science of medicine required apprenticeship to an experienced expert.

Building on this foundation, as well as leaping a century forward, medical teaching continued to proliferate its discourse. This discourse, like in every discipline, offers a way to communicate its understanding. In that aspect, it both creates and describes a world view.[2] Discourse has been expertly described as[3] "humans engaging in socially, culturally, and historically situated activities of conversation and practice—to decide on what is legitimate activity in any field." It is clear that educational overtures are incorporated in this discourse along with paradigm, theoretical, or even purely philosophical position shifts. In that context discourse is one of the first process that is associated with the learners and deals with them not as passive recipients but as active agents, builders in the learning process. Introducing facilitation instead of teaching brings agency to the learner and makes the whole educational process authentically more learner centric.

These radical or gradual shifts have brought significant changes in the understanding and implementation of fundamental aspects of medical education.[4] For educators, medical learning has evolved into far more than the acquisition of knowledge, attitudes, and skills. It has become, at its core, the construction of a professional identity, the change of one from a common individual to an expert professional. For the learners, contemporary education means that they need to move beyond the implementation and the dependence to rote knowledge and practice, while maintaining scientific rigor in their practice. It comes as a natural consequence that this directly reflects in curriculum, pedagogy, and even assessment.[5,6] These aspiring healthcare professionals have to master independent thinking, efficient problem solving and capacity to adapt their abilities to previously not encountered problems and tasks. In short, they must transform from an aspiring learner to a healthcare expert.

The coveted traits of "the expert" have been developed significantly, transferring these shifts to the objectives and directions of the medicinal training venture. Healthcare teachers today are educating the doctors of tomorrow for the experts that they shall become, looking to create individuals who are skillful, mindful, ready to self-criticize, and self-evaluate their actions as well as to keep learning all through their career. Acknowledgment of these objectives has led to activity on helping students create competency both in "how to learn," and in "what to learn." In that context, the Carnegie Foundation for Teaching and Learning's call for change of medical education[6] led to four noteworthy proposals for the change of healthcare education and training:

1. instructing and figuring out how to advance vocational integration;
2. advancing habits for inquiry and improvement;
3. individualizing learning, through standardized assessment; and
4. supporting the dynamic advancement of the expert identity.

Supporting this overall educational paradigm, allegories for learning have been used in reframing the mindset of the learner. Sfard[7] portrays two such metaphors: "acquisition" and "participation." In the "acquisition" representation, mastering knowledge is viewed as the procurement of learning, aptitudes, properties, qualities, and capabilities, as in one gets "products." Acquisition fortifies learning as an individual procedure. As is noted, this metaphor is so profoundly inserted in our reasoning that we barely saw it until the point when different such paradigms started to rise. The second analogy is that of "participation." This sees learning not as a something procured or accomplished. Rather, learning is in the participation and, as a process, it is seen as nonstop. While acquisition infers that information can be exchanged irrespectively of circumstances, participation tackles learning as inseparably attached to its unique situation and implanted in the social procedures there. It is also noted[7] that it is not to the students' advantage to embrace only one of these metaphors. Rather, it should be strived for educational methodologies to bolster the utilization of both are required.

This participatory, situated learning, has a place with those socio–cultural viewpoints that affirm it as dependent and inseparably linked to its specific situation and to the social relations and practices there; it is a transformative procedure that happens through engagement in the exercises of a group. Vygotsky,[8] depicted learning as happening through activity, interceded both by others and by social artifacts. Lave and Wenger[9] utilized the term "groups of training" to portray the exercises of a gathering of individuals who meet up in a quest for a common endeavor. They depict the part of the newcomer to the group as one of "honest to goodness fringe cooperation." In this procedure, newcomers or fledglings start at the outskirts of a group by watching and performing fundamental assignments. As they turn out to be more gifted, they move all the more midway in the group. Through investment, dynamic engagement, and accepting expanding accountability, the individual expect and gains the parts, abilities, standards, and estimations of the way of life and group. Further, as students are changed through cooperation in the group, their interest, reciprocally, changes the group.

Such situated learning sees the student as more than a spectator or imitator, as a dynamic member, gaining from and with all group members.[10] Lave and Wenger[11] likewise supportively recognize a showing educational module and a learning educational module. Learning educational modules comprises of situated avenues for improvement, in which the group transforms into the learning asset and learning happens from numerous points of view. Demonstrative educational modules, by comparison, are developed as guidelines for newcomers, and along these structures and constrains, pathways for learning and understanding.

Core precepts II: implicit experiential learning and intelligent educational group practice

It has been postulated[12,13] that learning and interest in practice are indivisible. They represent a co-development, emerging from the connections between the learning outcomes as they emerge in the work environment and from how people effectively interact with those outcomes. The work environment offers both human accomplices and different educational artifacts to explore; these connections between people in the social setting add to the person's ability to process and to internalize information. The student is a dynamic part in this collaborative environment. As such, working environments can influence learning by their preparation to draw in students and their interest.

Eraut's model of learning[14,15] at work likewise incorporates both social and individual perspectives. It depicts informal learning at work that happens through involvement and communication with partners. It portrays[14] implicit understanding as something that may occur without clear educating and in which the individual has no conscious effort for learning. He depicts inferred information as learning of settings and organizations, gained through a procedure of socialization, perception, acceptance, and cooperation. He, too, sees learning as logically arranged in a sequence of exercises and the social relations inside which these exercises are installed. Finally he observes social and individual instances of learning as reciprocal activities. In contrast to solitary educational information transfer these are both a social and an individual procedure.

Eraut's idea of inferred information and tacit learning has a specific remarkable quality for healthcare training. Tacit learning might happen when students watch or experience circumstances that test their qualities. The overcoming of these difficulties can prompt the maturation of qualities, qualifications, and a non-cerebral polishing of manual skills. In these, individuals even though they are uninformed of the gap that exists between their upheld view and practice, they nevertheless, tacitly cover it.[16]

The thought of learning through experience[17] has been generally acknowledged in medicinal training. Experiential learning, as portrayed by Boud et al.,[18] includes reflection on involvement with the objective of transforming knowledge into learning. Experiential learning reinforces typical learning; reflection is proposed to develop understanding and to investigate the more extensive settings of knowledge. Situated learning can complement experiential learning by encapsulating the exploration and engagement inside the group's standards, qualities, and exercises.

In that context, the coined terms intelligent learning and intelligent practice are indispensable to all learning points of view. Reflection enables knowledge to be effectively assimilated.[19] Reflection and intelligent practice[20] are themselves complex ideas. In spite of the fact that the literature progressively underpins reflection as a basic way to understanding and acclimatizing new ideas, contextualizing learning, and empowering execution change, its consolidation is challenging.[21–23] Reflective learning includes a core dissection of experience in order to comprehend its setting in order to internalize efficiently new discovering that have occurred there.[17,19] For the individual, reflection is identified with mindfulness, self-control, self-checking, and continuous learning. For the group, arranged learning provides setting and social context inside which to coordinate and make sense of experiences. At the point when reflection is attempted between and among individuals, it fuses the setting inside which the experience happened and opens doors for absorbing aggregate knowledge and qualities.

These features of tacit learning and intelligent practice beg the question: What are the exact manners in which educating and absorbance of content appear to be unique when seen from these points of view? Conceivably, three aspects of expanding the social measurements of learning can be seen, including approaches that amplify engagement, that boost group benefit and that expand, and internalize effective common group procedures to guarantee both individual and group learning.[4] In fact, collaborative learning is established upon the view that students have a crucial role in the group and that their learning and interest add to the group's and to their core development. It includes a reflective approach; it requires welcoming students into the group, providing a feeling of sincere help, as well as educational organization bolstering the individual's learning and limiting hindrances to cooperation through authoritative support.[24] Learners are effectively engaged with meaningful assignments that add to persistent care and exercises represent to students an opportunity to mirror their expanding abilities and responsibility.[9,12] Learners encounter the collaborations, qualities, difficulties, and procedures of the group.

Putting theory into practice; hurdles and advantages of implementation

Today, healthcare educators are under extensive societal weight and budgetary limitations to improve the nature of training and in that way the safety of therapeutic care. The idea of "learning by doing" has turned out to be unfeasible, especially when invasive strategies and risky procedures are required.[25] Furthermore, despite the best endeavors of instructors, a few techniques are so uncommon in clinical practice that they are almost impossible for clinicians to "see and do," not to mention teach.[26] Such restrictions have motivated them to look for strategies showing therapeutic information while increasing procedural experience. The maxim of "see one, do one, show one" is no longer feasible. Luckily, the previous decade has seen an explosion of devices accessible to improve medicinal training. Three driving instruments have developed to instruct and evaluate medicinal students: Web-based training, virtual reality (VR), and high quality patient cases reproduction.

There are several evident points of interest to Web-based instructive items. They offer improved capacity to asynchronous access to data on demand. Production of new or repurposed content is immediate. Sight and sound incorporated as experiential means, enrich the experience with photographs, videos, or sound documents. Also, Intelligent distance learning utilizes the Internet for webcasts, and live virtual discourses through both voice and text communication.[25]

VR innovation has additionally started to be utilized for medicinal applications. In only a brief span, an astonishing number of uses have been implemented for different fields inside medicine. VR test systems incorporate laparoscopic surgery, anatomy education, and decision making skills.[27,28] Currently, the lion's share of accessible VR applications have been intended for the surgical specialties because of the overwhelming need for support of doctors in those fields in critical medical situations. To date, almost each and every medical VR application has gotten an overwhelmingly positive subjective reaction.

Disregarding the experiential modality, high-fidelity virtual patient educational systems improve the productivity of learning in a safe environment, reliably enabling medicinal students to increase clinical experience without relying upon random encounters with actual patients. Such innovation has turned out to be very acknowledged and progressively requested via students over specialty lines.[29–32]. While full simulators account for essentially manual practice (e.g., flight simulator training in non-medical context teaches the manual nuances of flying a plane), immersive situations basically offer a chance to rehearse and think about basic analytic and reflective skills.[33] As the field of simulation develops, avatar-based simulations grow progressively experiential with procedural segments to provide both content and contextual authenticity. Such immersive simulations, enable people and groups to hone empathy, straight-up clinical skills and multitasking. By exploring core medical problems, medical abilities can be recognized and enhanced in conjunction with therapeutic skills even in the emergency medicine sector.[33,34] Due to the fact that fundamental medical correlations can be modeled by the simulators, practically any clinical situation can be created. The remarkable difficulties of an emergency treatment or a crisis situation, for instance, can be sensibly reproduced by on-screen computer controlled characters playing wandering patients, relatives, or specialists, every one of whom must be dealt with by the user's embodiment-avatar.[35]

Implementing experiential learning for health I; from Web based virtual patients to 3D virtual environments

The pivotal challenge in healthcare training was, and still is that of enormous content. Medical information doubled every 2 years after 1980s.[36] This required that instructors utilized technological innovation if just to deal with this enormous volume and essential nature of topics.[37] Overarching, such undertakings pointed toward universal access to clinical skill advancement instruments.[38] Web and Information, Communication Technologies developed into pivotal factors of healthcare training. Important among this bouquet of instruments is the Virtual Patient (VP). Formally characterized by the MedBiquitous Consortium as "interactive computer simulations of real-life clinical scenarios for the purpose of medical training, education, or assessment,"[39] it soon developed to a central methodology for medical training. With a formal MedBiquitous standard having been defined,[40] VPs were acknowledged through deployment improvements and online availability as the accepted standard for problem based learning healthcare training.[41,42] To encourage this across the board proliferation, several innovations developed. One of them, OpenLabyrinth (OLab)[42] is characterized in its manual as "… an open source online activity modelling system that allows users to build interactive 'game-informed' educational activities such as virtual patients, simulations, games, mazes and algorithms. It has been designed to be adaptable and simple to use while retaining a wealth of game-like features."[43] This pervasively available electronic apparatus requires negligible equipment and little to none PC proficiency for accessing and using it. This ubiquitous nature of this platform makes the entire undertaking of experiential Technology Enhanced Learning (TEL) to pivot around this straightforward and powerful educational content delivery method and makes them the initial stage of this current work's undertaking to plot the way of implementations in TEL toward ever more immersive, connected and in general, effective and efficient medicinal training.

From Web based VPs to virtual reality, mapping the implementation of such experiential means has to start from the repurposing of VPs across platforms and their standardization as learning assets. In this discourse, we shall touch briefly on the issue of web VP discoverability and instead we shall focus on the ways that a transfer of them has been achieved to MUVEs such Second Life and OpenSim.

From annotating VPs for easy discovery and repurposing [44] to the move of implementing virtual scenarios for healthcare training in experiential means the journey starts with multi user virtual Environments (MUVEs) such as Second Life and OpenSim. The endeavors taken for exchanging VP on various introduction stages comprise of the "cross-platform" repurposing of VPs from the Web to 3D multiuser virtual environments.

MUVEs have been utilized for implementing virtual patients for expanding impact and immediacy of a learning resource in a very "game-aware" demographic of today's higher education students.[45,46] MUVEs or "Virtual Worlds" have been defined as "A synchronous, persistent network of people, represented as avatars, facilitated by networked computers."[47] This definition incorporates in itself the instructive edges that this modality

Figure 7.1 *A dental VP in SL.* The Chat box displays automated narrative communication.

gives. Networking, and synchronous persistence is a collaboration multiplier while, the feeling of presence that a customizable human avatar of the user offers, makes for an immersive and prompt engaging experience.[48] These elements all are assistive toward the general educational objective. A standout among the most generally utilized MUVEs for educational content deployment has been Second Life (SL) and its open source partner OpenSim (OS). These platforms have wide and far applications in healthcare education (for a brief review of the field in SL and OS please c.f. Antoniou et al.[45]).

For the implementation details, a straightforward strategy that was followed shall be outlined. Cases can be easily transferred in SL or OS as "point-and-click adventures." Feedback to the user can be provided through automated chat messages (Fig. 7.1) or through short introductory presentations (Fig. 7.2). The learner guides the activity through either different decision "notecards" (text panels in SL/OS) or by clicking on actual objects of the world. In all cases, the VP decision trees are passed on to the SL/OS MUVE verbatim as they were authored in OLab.

In implementing the VPs in OS/SL, extensive use of their Scripting Language (OSSL / LSL) is required. All the proper Web assets (e.g., pictures) can be archived to an OS server. OSSL/LSL are event/state-based languages. Scripting uses events, for example, clicking, listening for chat messages, or creating things on the world. Utilizing these as triggers complex situations and activity feedback can be coded. Case parts can be implemented as events, activated as the user touched (clicked) bits of the world

Figure 7.2 *A mini presentation for teaching MUVE basics in a SL/VP.*

and overcame challenges through decision making in the branching narratives. Each user response would trigger new interaction instances driving the case to the next milestone state.

In order to merge the VP authoring and 3D asset use, a data modeling scheme should also be implemented. Such a data scheme should encourage repurposing of online VPs in a 3D virtual environment through standard serializations for the depiction of this VP as an OS point and click "adventure." This data modeling scheme should link MUVE front ends (OS/SL) to standardized back ends. The serialization of this data scheme can be a model JSON encapsulation for communication of the OSSL with the website pages. The objective of such data model is the use of it in a wide range of interactions while in the meantime staying open to expanding functionalities even to an alternate gamification setting. In that context, information serialization should separate between 3D environment aspects (e.g., object interaction node, notecard interaction node, etc.) and the VP relevant stuff (e.g., NodeText which is the content that portrays this node's narrative or LinkTexts that is the description of each decision). The following part outlines a model of such a data model.

Regarding the MUVE part of the data modeling, each OS 3D asset can be defined as the actual graphical asset, for example, a house, a chair, etc., as presented in the simulator. This is reposited in any available OS Asset Repurposing Archive and has, according to the OS Object data scheme, the necessary attributes for unique identification. These include the URL of its Grid, its unique identifier—OS—UUID, OS Grid location and size as well as a declared channel for use in chat communication of the VP engine with the asset. Geometry Objects are hierarchically denoted as Assets and Environments in order to be able to establish prerequisite relations in their repurposing and re-use (e.g., a piece of pottery is an asset that belongs to the Environment "Garden"). That way content creators can meaningfully interpret relationships between Assets. For example, if a user was to repurpose a VP and chooses "Dentist's Office" environment, there should be clarity about which assets of the database exist in the same OSGrid as well as which are part of this generically defined "Dentist's Office." This OS Object data scheme links to OLab's data scheme so that meaningful state/link pairings can be assigned to each instance of use for each Asset in the MUVE. Each core entity type of OpenLabyrinth (node, link, or case) has equivalent there. For example, there are classes of Labyrinth, Node, and Link, which represent the same structure found in the original virtual patient. The interaction mapping between these OpenLabyrinth entities and the OS Assets is realized through another set of relevant classes. NodeNavigation is a wrapper class of the typical OLab Node class. Instances of this class are when users navigate new nodes through links. Such a NodeNavigation instance requires an InteractionTriggerCollection that depicts all the links of the current node, thus enabling multi-type interactivity. Depending on the properties of the node, this InteractionTriggerCollection can be implemented either as a NoteCardOptionCollection or as a LinkBoxCollection to address the specific scripting implementation of either a Notecard or a clickable objects' nodes. The first one needs to be attached in an OS asset so it triggers many NoteCardOption instances to show as message options in the OS Notecard. On the other hand, in order to create the clickable object interactivity trigger boxes in the environment should contain LinkBoxes class instances that are grouped in a LinkBoxCollection for each Node instance. For an even more detailed discourse as well as for the technical details of implementing such a model and front end in OpenSim, the reader can refer to the detailed technical paper describing fully the implementation.[49]

Implementing experiential learning for health II; virtual, mixed reality and the educational living lab

If repurposing of VPs in MUVEs is the logical first step in implementing experiential digital health learning, then immersive interfaces through contemporary technologies is the straightforward evolution. With the advent of Virtual Reality (VR) and the even more relevant, learning-wise, Augmented Reality (AR) the educational space of Virtual labs was re-invigorated.

Virtual labs use the qualities of electronic simulation and recreation along with a large group of other instructional advancements (video etc.) to supplant a genuine collaborating lab experience. A Virtual lab, for instance could comprise of a large group of advanced simulation connected with forum messages, video presentations, or even other experiential simulations.[50] This sort immersive integrative interactions enable the learners to pace their learning procedure by repeating content, getting to it at off hours and generally keeping up the activity in their learning procedure. Past that, preparation and sharpening of such lab aptitudes covers a portion of the central shortcomings of contemporary medical educational programs. Essential information in medicine, for example, biology and genetics runs hand in hand with hands on capacities in a lab setting. With the present status of cutting edge diagnostic methods, for example, genome tests, it is crucial that doctors have a solid scientific foundation.[51] Nevertheless, novel lab techniques are not available for hands on training given economic and other overall logistics factors.[52,53] These limitations point medical instruction to theoretical understanding leaving future doctors lacking in certifiable lab and clinical aptitudes.[53]

An approach that has been taken to convey genuine abilities to training (though so far for the most part in the sciences) has been the consolidation of rising Virtual and Augmented Reality (VR/AR) technology advancements. It has been exhibited that AR innovation can essentially multiply instructive effect of the learning content and in this way significantly influence the instructive result.[54] Cases have included world investigation at an extremely experiential level,[55] or encountering science and material science ideas in a way that is difficult to accomplish without it. Envisioning chemistry or physics interactions[56] or the ideas of attractive magnetic fields or wind stream[57,58] is something that can't be accomplished outside an AR domain with this kind of immediacy. This sort of immersiveness is additionally responsible for spurring learners to additionally investigate topics on their own and intrinsically figure out how to avoid conceptual mistakes.[59]

Practically, the way to implement an immersive AR lab that can facilitate VP integration consists of a real world space, a simple room that shall be enriched with a series of AR image targets. Given the premise of AR such targets act as triggers for any device running the relevant software through which to present the digital content (3D models, animation, etc.) as well as through which all interactions are realized. To make the experience more immersive, the user usually wears an inexpensive digital eyewear headset

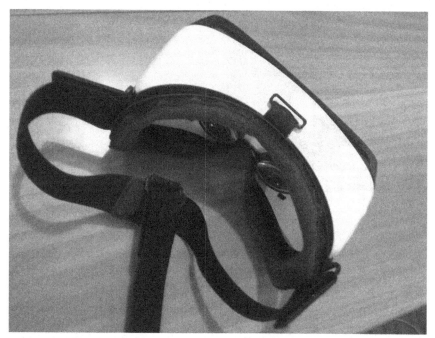

Figure 7.3 *Inexpensive AR headset utilizing mobile devices as AR application deployment platform.*

that incorporates mobile devices (Fig. 7.3) as the deployment platform for the digital content of the VP scenario.

Regarding the interaction logic of a VP in such an environment, this follows the same stateful approach as has been described earlier. The game logic keeps track of the progress across the nodes of the VP episode as it unfolds in the mixed reality environment and displays content and feedback as the user progresses through the VP nodes. With the current advent of visual development platforms and the lack of significant IT architectural overhead the choice of development ecosystem should be on that allows for easy and extensible development. Custom game engines or standalone 3D solutions are not warranted for effectively translating the educational narrative to the envisioned gamified mixed reality environment. Given these characteristics, the application front end is created utilizing Unity 3D and the "Vuforia" AR/digital eyewear platform (Fig. 7.4). This environment provides the usability essentials for quick implementation, while keeping up an adequate level of adaptability for the kind and volume of content that is expected for such an implementational effort.

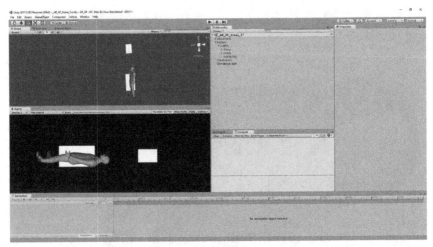

Figure 7.4 *Unity development environment with the Vuforia extensions installed.* Placeholder 3D assets overlaid over template AR target images.

An exemplar case that was developed and evaluated was a cardiology vignette.[28] For this situation, the user needed to treat a patient who, amidst a normal cardiology residency, experienced a myocardial infraction. The vignette that was utilized comprised of the underlying patient examination up to the point of the principal manifestations of the myocardial infraction. This vignette was an apt port from the Web for our objective (repurposing VPs in an educational lab condition), since it included interaction in several places of the lab's space. To help this connection the execution used image targets on the walls of the lab and Vuforia's virtual button innovation (even though any event based AR interactivity technology would do). Through such interaction modalities, the user interacted with blended reality objects (e.g., a genuine electrocardiogram picture, augmented with 3D resources, and virtual buttons) in order to get feedback from the environment. That way through tactile interactions, users would get acclimated to the intuitiveness in introduction of a VP in the blended reality educational space.

Toward mixed reality educational living labs I; supporting repurposing of Web assets to 3D immersive learning environments

Through the previous discourse it becomes apparent that technologies for immersive VR/AR content in VP medical experiential education are becoming rather mature and available for quick integration of content.

This creates the problem of streamlining content transfer and efficient re-purposing from Web based resources to these novel modalities.

Virtual Learning Environment (VLE) have advanced fundamentally since their initiation. From single online repositories VLEs are presently imagined as community oriented, decentralized, instructive action direct lights that, expertly, participate toward particular learning objectives. This worldview of decentralization adjusts well to repurposing endeavors. With the immersion potential, the capacity toward collaboration and engagement of a mixer reality environment, the consideration of this instructive medium in the medical education arsenal shows up as a key development. The primary hurdles to such advancement are both the development of the technology, the implementation of the infrastructure, and the production of a non-trivial measure of content in this medium. With the previously mentioned emerging platforms for deployment of VR/AR content and the proliferation of m-infrastructure (mobile phones and tablets), the first two issues are becoming increasingly irrelevant. Thus, the repurposing issue becomes central. With VPs as an implementation being around for around 40 years[60] and web organizations and authoring platforms like OpenLabyrinth having matured,[42] electronic VPs can be viewed as a standard apparatus in medical instruction. Subsequently, streamlining exchange of such cases to a 3D environment can provide huge content support in MUVE/VR/AR VPs. Besides, the reusability of 3D resources in a large number of cases (stored in a back end database scheme) lighten to a critical degree the requirement for large volumes of 3D content creation.

Content creation and repurposing endeavors in experiential 3D environments obviously are not novel. Particularly in fields heavy with aesthetic assets at their core, for example, cultural heritage, there have been undertakings of transferring real world content to 3D environments for education or exposure of more people to it (e.g., portrayals of masterpieces, landmarks, and so forth).[61,62] In that field, also, there have been achievements in the utilization of semantic annotations and enrichment of content toward streamlining the process of exchange from the real to the virtual space.[63] Metadata, embedded from wikis and established semantic namespaces, along with geospace data, have been effectively utilized for offering constant updating of 3D environments inspired from real world destinations.[64]

Moving to healthcare training needs to exist, it is sure that the fuse of online virtual patients as an educational modality isn't an oddity; there have been investigations of their effect and viability as course materials.[65,66] Moreover, the progress from the Web to the virtual space needs to offer a solid validation of pedagogic value of this transfer of content to the 3D immersive modality. In that front, past works have advanced[45] best practices and guidelines with respect to fitting and educationally adapt repurposing of the virtual patient content from the Web to a 3D MUVE. In any case, moving to specifics in the previously mentioned endeavors, two axes of research emerge. One is the overcoming of the technical hurdles for transfer of the VP logic and presentation to the 3D environment (MUVE/VR/ AR). The other is the facilitation of automated content repurposing in this new deployment platform by semantic annotation and enrichment. To put it plainly, there is a need both for a simulation engine to present VPs in the 3D modality, but also there is a need for a streamlined way of transferring web VP components (links/nodes/cases) in the MUVE/VR/AR spaces. Usage of relevant web tools such as web services and architecturally sound back-ends should enable specialists and other content suppliers to afford such expedient repurposing of VPs into MUVE/VR/AR.

Toward mixed reality educational living labs II; co-creative immersive educational spaces

The endeavors mentioned earlier are not just an exercise in streamlined data modeling and serious game development. It directly reflects the necessities of an integrative educational environment utilizing experiential modalities like VR and AR. It has been exhibited that the utilization of AR innovations decidedly influences students' learning and abilities by experientially and intuitively immersing them into the topic.[58,67–70] In the current technological state, AR applications can be supported by cell phones. Consequently, instructors wishing to utilize AR environments have a ubiquitous stage for prompt access to a mixed reality learning environment accessible in both institutional and individual setting. It has been asserted that it is better for students to work not in a totally virtual reality lab, thus the mixed reality learning environment can be considered as the ideal arrangement.[59] Studies also have exhibited that mixed reality labs have had positive effect in the expediency with which students finish lab experiments.[67] The reason behind this distinction is the capacity of students to internalize underlying basic principles in their reasoning quicker through experiential immersion. This same mental process was additionally found to be a factor in enhancing student's performance in especially challenging for him topics. In short, motivation and engagement of the student increases altogether[67] due to experiential immersion. Thus the inclusion of digital content in real environments, as it appears, offers to learners/participants, the same versatility and impact as the contemporary research environments Living Labs (LL) have provided without digital enhancement.

A LL is an environment either housing or emulating the housing of people and technology (teachers/facilitators and sensors/actuators), in an non-invasive but ecologically valid setting that supports core activities relevant to the environment and the conduct of validated research on these.[71] For example, a staged sitting room area that contains a setup of cameras and microphones can allow people inside it to emulate their daily routines while researchers can mine invaluable data for factors ranging from ergonomy to personal health issues. More formally, LLs have been characterized as "a user-centric innovation milieu built on every-day practice and research, with an approach that facilitates user influence in open and distributed innovation processes engaging all relevant partners in real-life contexts, aiming to create sustainable values."[72] From this definition it follows that LLs can empower organizations, specialists, scientists, and clients to work together toward development, inquire about, co-create, and showcase relevant topics in authentic real-life situations.[72] Existing writing offers calculated exchanges about co-creation in LLs. Research, from focus groups, interviews and observations converge to six basic components for cultivating co-creation as an impact multiplier in LLs[73]: Engagement, Relationship Management, Atmosphere, Operating Principle, Design Layout, and Data Collection Approach.

That mix's core quality, Engagement, is also an essential objective for the successful deployment of a mixed reality educational environment in an authentic setting. The second factor, Relationship Management that alludes to the functional relationship between the different actors in a collaborative endeavor[74] is directly analogous to the collaboration necessities that exist in a contemporary educational setting. The third factor, Atmosphere, is precisely the immersion as assistance and empowering influence that is the very core for the deployment of a mixed reality environment. Operating Principles, in LLs, as an arrangement of overall rules for ideal utilization of the ecologically valid environment in its unique circumstance, is specifically applicable to the particular instructive ideal models (PBL, educational programs streamlining) that facilitate the mixed reality educational environment. Besides, Design Layout (making the environment as similar to a realistic one even though it contains a research-sensor vector) is as an essential factor for encouraging co-creation in LLs but also the fundamental edge of the mixed reality 3D environment in comparison to other forms of digital healthcare education. Finally, the Data Collection Approach of a LL domain includes both the sensors' design and the information gathering methodology. In that regard, a mixed reality educational environment caters for data collection with intrinsic provisions for application integration of user and overall learning analytics for extracting crucial research outcomes and teaching aids, often in real-time. Given these immediate analogies, it becomes clear that immersive experiential mixed reality's natural next evolutionary step is a Co-creative Educational Living Lab.

Toward mixed reality educational living labs III; visualizing the mixed reality educational living lab

Given the limited number of students and personnel at any given site, it is troublesome for singular projects to create the measurable power required to legitimately evaluate the overall adequacy of developing educational innovations. It is, thus, basic for educational specialists to collaborate beyond localization limitations. What emerges from this brief discourse about both the conceptual base, the hurdles and the technicalities of implementation in Technology enhanced learning is that there is the necessity of thoroughly planned endeavors for the advancement, appraisal, and execution of emerging educational innovations. This is the only way to guarantee the shift in medical training thus improving the performance in multiple medical specialties. In that context, virtual patients, VR, and the World Wide Web empower medical learners, both undergraduates and vocational trainees to experience a plethora of cases despite of location, patient inflow, and time of training in order to both learn and become themselves capable professionals in order to train the next generation of their future peers.[25]

As has been presented earlier, this crucial outcome passes from the endeavor of implementing a mixed reality educational LL. In that context, it requires a disruptive innovation not in technology but in the conceptual framework of implementing Digital Learning. It requires a view of immersive experiential learning through the lens of game design principles. Scientifically effective narrative is only one of the components for such an implementation. A merger of all spatial degrees of freedom (3D animation and actual user movement) is required with natural interfaces (voice, gesture) to create an environment of immersion, engagement, and educational impact that shall allow the learner to feel as an actor and not a receiver in the educational experience.

An actual mixed reality Living Lab aims to utilize the previously described implementations not only to demonstrate static cases to the learner. It aims to integrate cases in high granularity, enabling the repurposing at the node level. With that capacity, expert systems functionality can create new emergent cases. Combined with analytics integrated from each user's previous interactions, such a system would enable the user to become even without effort a co-creator of digital healthcare resources that are in turn going to be discoverable and repurposable for massive proliferation of experiential immersive medical education content.

References

1. Flaxner A. Medical education in the United States and Canada. A Report of the Carnegie Foundation for the Advancement of Teaching. *Carnegie Found Adv Teaching, Bull Number Four*. Available from: http://www.scielosp.org/pdf/bwho/v80n7/a12v80n7.pdf; 1910 [cited 2017 Oct 11].
2. Wetherell M, Yates S, Taylor S, Open University. *Discourse theory and practice: a reader*. SAGE; 2001; p. 406. Available from: https://books.google.gr/books?hl=el&lr=&id=uDFvN1M3_sMC&oi=fnd&pg=PA1&ots=zOXNIIckzg&sig=UzHTge3g9wFfNpSeg6XTVgTK1DI&redir_esc=y#v=onepage&q&f=false.
3. Bleakley A. Curriculum as conversation. *Adv Heal Sci Educ* 2009;**14**(3):297–301 [Available from: http://link.springer.com/10.1007/s10459-009-9170-6.
4. Mann KV. Theoretical perspectives in medical education: past experience and future possibilities. *Med Educ* 2011;**45**(1):60–8 Available from: http://doi.wiley.com/10.1111/j.1365-2923.2010.03757.x.
5. Bereiter C, Scardamalia M. Surpassing ourselves: an inquiry into the nature and implications of expertise. *Surpassing ourselves An Inq into Nat Implic Expert* 1993;279. Available from: http://ikit.org/fulltext/1993surpassing/preface.pdf.
6. Irby DM, Cooke M, O'Brien BC. Calls for reform of medical education by the Carnegie Foundation for the advancement of teaching: 1910 and 2010. *Acad Med* 2010;**85**(2):220–7 [cited 2017 Oct 11]. Available from: http://content.wkhealth.com/linkback/openurl?sid=WKPTLP:landingpage&an=00001888-201002000-00018.
7. Sfard A. On two metaphors for learning and the dangers of choosing just one. *Educ Res* 1998;**27**(2):4–13 [cited 2017 Oct 31]. Available from: http://edr.sagepub.com/cgi/doi/10.3102/0013189X027002004.
8. Vygotskij LS, Cole ME. *Mind in society: the development of higher psychological processes [Internet]* Cambridge, MA: Harvard University Press; 1978. p. 159 DOI: 10.2307/j.ctvjf9vz4.
9. Lave J, Wenger E. *Situated learning: legitimate peripheral participation*. Cambridge, UK: Cambridge University Press; 1991. p. 138.
10. Egan T, Jaye C. Communities of clinical practice: the social organization of clinical learning. *Heal An Interdiscip J Soc Study Heal Illn Med* 2009;**13**(1):107–25 [cited 2017 Nov 11] Available from: http://journals.sagepub.com/doi/10.1177/1363459308097363.
11. Lave J, Wenger E. Legitimate peripheral participation in communities of practice. *Supporting Lifelong Learning I* 2002;111–26 [cited 2017 Nov 11] Available from: https://scholar.google.gr/scholar?cluster=1848085332532862798&hl=el&as_sdt=0,5.
12. Billett S. Learning through work: workplace affordances and individual engagement. *J Work Learn* 2001;**13**(5):209–14 [cited 2017 Nov 11] Available from: http://www.emeraldinsight.com/doi/10.1108/EUM0000000005548.

13. Billett S. Workplace participatory practices. *J Work Learn* 2004;**16**(6):312–24 [cited 2017 Nov 11] Available from: http://www.emeraldinsight.com/doi/10.1108/13665620410550295.

14. Eraut M. Non-formal learning and tacit knowledge in professional work. *Br J Educ Psychol* 2000;**70**(1):113–36 [cited 2017 Nov 11] Available from: http://doi.wiley.com/10.1348/000709900158001.

15. Eraut M. Learning from other people in the workplace. *Oxford Rev Educ* 2007;**33**(4):403–22 [cited 2017 Nov 11] Available from: http://www.tandfonline.com/doi/abs/10.1080/03054980701425706.

16. Coulehan J, Williams PC. Vanquishing virtue: the impact of medical education. *Acad Med* 2001;**76**(6):598–605 [cited 2017 Nov 11] Available from: http://www.ncbi.nlm.nih.gov/pubmed/11401802.

17. Kolb DA. Experiential learning: experience as the source of learning and development. 390p.

18. Boud D, Keogh R, Walker D. Reflection: turning experience into learning. *Learning* 1985;170 [cited 2017 Nov 11] Available from: https://books. google. gr/books?hl=el&lr=&id=XuBEAQAAQBAJ&oi=fnd&pg=PP1&dq=Reflection:+Turning+Experience+into+Learning. &ots=TuXn3Wqh_S&sig=PSVbKN5_INGvaB3HTtgdtTn-tF4&redir_esc=y#v=onepage&q=Reflection%3A Turning Experience into Learning. &f=false.

19. Moon J. *Reflection in learning and professional development: theory and practice*Kogan Page; 2000. p. 229 [cited 2017 Nov 11] Available from: https://books. google. gr/books?hl=el&lr=&id=8y0LwQxZUf4C&oi=fnd&pg=PR2&dq=Reflection+in+Learning+and+Professional+Development. &ots=1sXvvzPTlY&sig=R0LuM9bOvOLT1agTlbIij82Xvkc&redir_esc=y#v=onepage&q=Reflection in Learning and Professional Development. &f=false.

20. Schön DA. *Educating the reflective practitioner: toward a new design for teaching and learning in the professions*San Francisco: Jossey-Bass; 1987 [cited 2017 Nov 11] Available from: http://psycnet.apa.org/record/1987-97655-000.

21. Mann K, Gordon J, MacLeod A. Reflection and reflective practice in health professions education: a systematic review. *Adv Heal Sci Educ* 2009;**14**(4):595–621 [cited 2017 Nov 11] Available from: http://link.springer.com/10.1007/s10459-007-9090-2.

22. Boud D, Walker D. Promoting reflection in professional courses: the challenge of context. *Stud High Educ* 1998;**23**(2):191–206 [cited 2017 Nov 11] Available from: http://www.tandfonline.com/doi/abs/10.1080/03075079812331380384.

23. Grant A, Kinnersley P, Metcalf E, Pill R, Houston H. Students' views of reflective learning techniques: an efficacy study at a UK medical school. *Med Educ* 2006;**40**(4):379–88 [cited 2017 Nov 11] Available from: http://doi.wiley.com/10.1111/j.1365-2929.2006.02415.x.

24. Dornan T, Boshuizen H, King N, Scherpbier A. Experience-based learning: a model linking the processes and outcomes of medical students' workplace learning. *Med Educ* 2007;**41**(1):84–91 [cited 2017 Nov 11] Available from: http://doi.wiley.com/10.1111/j.1365-2929.2006.02652.x.

25. Vozenilek J, Huff JS, Reznek M, Gordon JA. See one, do one, teach one: advanced technology in medical education. In: *Academic Emergency Medicine* [Internet]. 2004; p. 1149–54. [cited 2017 Oct 3] Available from: http://doi.wiley.com/10.1197/j.aem.2004.08.003.

26. Hayden SR, Panacek EA. Procedural competency in emergency medicine: the current range of resident experience. *Acad Emerg Med* 1999;**6**(7):728–35 [cited 2017 Nov 11] Available from: http://doi.wiley.com/10.1111/j.1553-2712.1999.tb00444.x.

27. Reznek M. Virtual reality and simulation: training the future emergency physician. *Acad Emerg Med* 2002;**9**(1):78–87 [cited 2017 Nov 11] Available from: http://doi.wiley.com/10.1197/aemj.9.1.78.

28. Antoniou PEPEP, Dafli E, Arfaras G, Bamidis PDPPD. Versatile mixed reality medical educational spaces; requirement analysis from expert users. In: *Personal and Ubiquitous Computing* [Internet]. 2017; p. 1–10. [cited 2017 Sep 5] Available from: https://link. springer.com/article/10.1007/s00779-017-1074-5.

29. Halamek LP, Kaegi DM, Gaba DM, Sowb YA, Smith BC, Smith BE, et al. Time for a new paradigm in pediatric medical education: teaching neonatal resuscitation in a simulated delivery room environment. *Pediatrics* 2000;**106**(4):E45 [cited 2017 Nov 11] Available from: http://www.ncbi.nlm.nih.gov/pubmed/11015540.

30. Bond WF, Kostenbader M, McCarthy JF. Prehospital and hospital-based health care providers' experience with a human patient simulator. *Prehosp Emerg Care* [Internet]. [cited 2017 Nov 11] 2001;**5**(3):284-7. Available from: http://www.ncbi.nlm.nih.gov/ pubmed/11446544.

31. Dawson SL, Cotin S, Meglan D, Shaffer DW, Ferrell MA. Designing a computer-based simulator for interventional cardiology training. *Catheter Cardiovasc Interv* 2000;**51**(4):522-7 [cited 2017 Nov 11] Available from: http://www.ncbi.nlm.nih.gov/ pubmed/11108693.

32. Gordon JA, Wilkerson WM, Shaffer DW, Armstrong EG. 'Practicing' medicine without risk: students' and educators' responses to high-fidelity patient simulation. *Acad Med* 2001;**76**(5):469-72 [cited 2017 Nov 11] Available from: http://www.ncbi.nlm.nih.gov/ pubmed/11346525.

33. Bond WF, Deitrick LM, Arnold DC, Kostenbader M, Barr GC, Kimmel SR, et al. Using simulation to instruct emergency medicine residents in cognitive forcing strategies. *Acad Med* 2004;**79**(5):438-46 [cited 2017 Nov 11] Available from: http://www.ncbi.nlm.nih. gov/pubmed/15107283.

34. Gaba, David M, Gaba DM, Howard SK, Fish KJ, Smith BE, Sowb YA. Simulation-based training in anesthesia crisis resource management (ACRM): a decade of experience. Crew Resource Management: In: *Critical Essays ISO 690.* 2017.

35. Kobayashi L, Shapiro MJ, Suner S, Williams KA. Disaster medicine: the potential role of high fidelity medical simulation for mass casualty incident training. *Med Heal* 2003;**86**(7):196.

36. Schittek M, Mattheos N, Lyon HC, Attström R. Computer assisted learning. A review. *J Dent Educ* 2001;**5**(3):93–100 [cited 2016 Jul 2] Available from: http://www.ncbi.nlm. nih.gov/pubmed/11520331.

37. Fry H, Ketteridge S, Marshall S. A handbook for teaching and learning in higher educa-tion: enhancing academic practice. In 1999 [cited 2016 Jul 2]. Available from: http:// books.google.co.uk/books?hl=en&lr=&id=5Rp9AgAAQBAJ&oi=fnd&pg=PP1&dq =a+handbook+for+teaching+and+learning+for+higher+education+heather+fry&o ts=_cfCpdr_Xk&sig=h7meUYcYzd4UxhHA1wlRyg-Ip1U#v=onepage&q=a hand-book for teaching and learning for higher edu.

38. Downes S. Distance Educators Before the River Styx. *Technol Source*, 2001. Available from: http://technologysource.org/article/distance_educators_before_the_river_styx/.

39. Ellaway R, Candler C, Greene P, Smothers V. An architectural model for MedBiquitous virtual patients. *MedBiquitous* 2006;**6**:1–15 Available from: http://groups. medbiq. org/ medbiq/display/VPWG/MedBiquitous+Virtual+Patient+Architecture.

40. Ellaway R, Poulton T, Fors U, McGee JB, Albright S. Building a virtual patient com-mons. *Med Teach* 2008;**30**(2):170–4 [cited 2014 Jan 31] Available from: http://www. tandfonline.com/doi/full/10.1080/01421590701874074.

41. Poulton T, Balasubramaniam C, Poulton Chara T. Virtual patients: a year of change. *Med Teach* 2011;**33**(11):933–7 [cited 2016 Jul 2] Available from: 10.3109/0142159X. 2011.613501%5Cnhttp://www.proxy.its.virginia.edu/login?url=http://search.ebscohost. com/login.aspx?direct=true&db=a2h&AN=66792420&site=ehost-live%5Cnhttp:// www.proxy.its.virginia.edu/login?url=http://search.ebscohost.com/login.aspx?direct=.

42. Ellaway RH. OpenLabyrinth: an abstract pathway-based serious game engine for professional education. In: *2010 Fifth international conference on digital information management (ICDIM)* [Internet]. IEEE; 2010; p. 490–5. Available from: http://ieeexplore.ieee.org/document/5664241/.
43. *OpenLabyrinth User Guide* [Internet]. [cited 2016 Jul 2]. Available from: http://openlabyrinth.ca/wp-content/uploads/2013/04/OpenLabyrinth-v3-User-Guide.docx.
44. Dafli E, Antoniou P, Ioannidis L, Dombros N, Topps D, Bamidis PD. Virtual patients on the semantic web: a proof-of-application study. *J Med Internet Res* 2015;**17**(1):e16 Available from: http://www.jmir.org/2015/1/e16/.
45. Antoniou PEPE, Athanasopoulou CACA, Dafli E, Bamidis PDPD. Exploring design requirements for repurposing dental virtual patients from the web to second life: a focus group study. *J Med Internet Res* 2014;**16**(6):1–19.
46. Papadopoulos L, Pentzou AE, Louloudiadis K, Tsiatsos TK. Design and evaluation of a simulation for pediatric dentistry in virtual worlds. *J Med Internet Res* 2013;**15**(11):e240 [cited 2014 Feb 13] Available from: http://www.pubmedcentral.nih.gov/articlerender.fcgi?artid=3841347&tool=pmcentrez&rendertype=abstract.
47. Bell MW. Toward a definition of virtual worlds. *J Virtual Worlds Res* 2008;**1**(1):1–5.
48. Boulos MNK, Hetherington L, Wheeler S. Second Life: an overview of the potential of 3-D virtual worlds in medical and health education. *Health Info Libr J* 2007;**24**(4):233–45.
49. Antoniou PE, Ioannidis L, Bamidis PD. OSCase: data schemes, architecture and implementation details of virtual patient repurposing in multi user virtual environments. *EAI Endorsed Trans Futur Intell Educ Environ* 2016;**2**(6):151523 [cited 2016 Jul 16] Available from: http://eudl.eu/doi/10.4108/eai.27-6-2016.151523.
50. Scheckler RK. Virtual labs: a substitute for traditional labs? *Int J Dev Biol* 2003;**47**(2–3):231–6 [cited 2016 Jul 3] Available from: http://www.ncbi.nlm.nih.gov/pubmed/12705675.
51. Makransky G, Bonde MT, Wulff JSG, Wandall J, Hood M, Creed PA, et al. Simulation based virtual learning environment in medical genetics counseling: an example of bridging the gap between theory and practice in medical education. *BMC Med Educ* 2016;**16**(1):98 [cited 2016 Jul 3] Available from: http://bmcmededuc.biomedcentral.com/articles/10.1186/s12909-016-0620-6.
52. de Jong T, Linn MC, Zacharia ZC. Physical and virtual laboratories in science and engineering education. *Science* 2013;**340**(6130):305–8 [cited 2016 Jul 3] Available from: http://www.ncbi.nlm.nih.gov/pubmed/23599479.
53. Weller JM. Simulation in undergraduate medical education: bridging the gap between theory and practice. *Med Educ* 2004;**38**(1):32–8 [cited 2016 Jul 3] Available from: http://www.ncbi.nlm.nih.gov/pubmed/14962024.
54. Chiu JL, DeJaegher CJ, Chao J. The effects of augmented virtual science laboratories on middle school students' understanding of gas properties. *Comput Educ* 2015;**85**:59–73.
55. Dede C. Immersive interfaces for engagement and learning. *Science* 2009;**323**(5910):66–9 [cited 2016 Jul 3] Available from: http://www.ncbi.nlm.nih.gov/pubmed/19119219.
56. Klopfer E, Squire K. Environmental Detectives—the development of an augmented reality platform for environmental simulations. *Educ Technol Res Dev* 2008;**56**(2):203–28 [cited 2016 Jul 3] Available from: http://link.springer.com/10.1007/s11423-007-9037-6.
57. Dunleavy M, Dede C, Mitchell R. Affordances and limitations of immersive participatory augmented reality simulations for teaching and learning. *J Sci Educ Technol* 2009;**18**(1):7–22 [cited 2016 Jul 3] Available from: http://link.springer.com/10.1007/s10956-008-9119-1.
58. Wu HH-K, Lee SW-Y, Chang H-Y.H., Liang J-CJJ-C, Wen S, Lee-Yu, et al. Current status, opportunities and challenges of augmented reality in education. *Comput Educ* 2013;**62**:41–9 [cited 2016 Jul 1] Available from: https://ac.els-cdn.com/S0360131512002527/1-s2.0-S0360131512002527-main.pdf?_tid=73bb82ac-a842-11e7-b3c3-00000aab0f02&acdnat=1507039083_3021d9a01b3b8f18a7bcf48a62d81ce6.

59. Olympiou G, Zacharia ZC. Blending physical and virtual manipulatives: an effort to improve students' conceptual understanding through science laboratory experimentation. *Sci Educ* 2012;**96**(1):21–47 [cited 2016 Jul 1] Available from: http://doi.wiley.com/10.1002/sce.20463.

60. Cook DA, Triola MM. Virtual patients: a critical literature review and proposed next steps. *Med Educ* 2009;**43**(4):303–11 [cited 2014 Jan 23] Available from: http://www.ncbi.nlm.nih.gov/pubmed/19335571.

61. Meli M. *Knowledge management: a new challenge for science museums*; 2003.

62. Marcos G, Eskudero H, Lamsfus C, Linaza MT. *Data retrieval from a cultural knowledge database.* Available from: http://citeseerx.ist.psu.edu/viewdoc/download?doi=10.1.1.390.9705&rep=rep1&type=pdf.

63. Mrissa M, Dietze S, Thiran P, Ghedira C, Benslimane D, Maamar Z. Context-based semantic mediation in web service communities. *Weaving services and people on the World Wide Web*Berlin Heidelberg: Springer; 2009. p. 49–66 [cited 2017 Nov 22] Available from: http://link.springer.com/10.1007/978-3-642-00570-1_3.

64. Dunwell I, Petridis P, Protopsaltis A, Freitas SDe. *Automating content generation for large-scale virtual learning environments using semantic Web services.*

65. Edelbring S, Broström O, Henriksson P, Vassiliou D, Spaak J, Dahlgren LO, et al. Integrating virtual patients into courses: follow-up seminars and perceived benefit. *Med Educ* 2012;**46**(4):417–25 [cited 2014 Jan 23] Available from: http://www.ncbi.nlm.nih.gov/pubmed/22429178.

66. Zary N, Johnson G, Fors U. Web-based virtual patients in dentistry: factors influencing the use of cases in the Web-SP system. *Eur J Dent Educ* 2009;**13**(1):2–9.

67. Akçayir M, Akçayir G, Pektas, HM, Ocak MA. Augmented reality in science laboratories: the effects of augmented reality on university students' laboratory skills and attitudes toward science laboratories. *Comput Human Behav* 2016;**57**:334–42.

68. Cai S, Wang X, Chiang F-K. A case study of augmented reality simulation system application in a chemistry course. *Comput Human Behav* 2014;**37**:31–40.

69. Chen C-M, Tsai Y-N. Interactive augmented reality system for enhancing library instruction in elementary schools. *Comput Educ* 2012;**59**(2):638–52.

70. Fulantelli G, Taibi D, Arrigo M. A framework to support educational decision making in mobile learning. *Comput Human Behav* 2015;**47**:50–9.

71. Chin J, Callaghan V. Educational living labs: a novel Internet-of-Things based approach to teaching and research. In: *2013 ninth international conference on intelligent environments.* IEEE; 2013. p. 92–9. [cited 2016 Jul 1] Available from: http://ieeexplore.ieee.org/lpdocs/epic03/wrapper.htm?arnumber=6597796.

72. Bergvall-Kåreborn B, Ståhlbröst A. Living lab: an open and citizen-centric approach for innovation. *Int J Innov Reg Dev* 2009;**1**(4):356–70.

73. Greve K, Martinez V, Jonas J, Neely A, Möslein K. *Facilitating co-creation in living labs: the JOSEPHS study.* 2016.

74. Ogunnaike OO, Borishade TT, Jeje OE. Customer relationship management approach and student satisfaction in higher education marketing. *J Compet* 2014;**6**(3):49–62 [cited 2016 Jul 1] Available from: http://www.cjournal.cz/index.php?hid=clanek&cid=175.

CHAPTER EIGHT

Artificially intelligent chatbots for health professions education

Janet Corral
University of Arizona Tucson College of Medicine, Tucson, AZ, United States

Chapter outline

What are chatbots?

Chatbots are "machine conversation system[s] [that] interact with human users via natural conversational language,"[1] (p.489). Chatbots offer computer-programmed conversations that focus answering questions from humans, as well as dialoguing with humans. The basic model of chatbots is a question-and-answer format, where direct questions are programmed with set answers that the computer retrieves and provides to the human through a web-based window or interface. This static approach may make a chatbot easier to program in that one must only develop the content for the question-and-answer conversation. In the last decade, chatbots have evolved from simple text-based forms, where users type in questions and receive answers, to audio-based chatbots, which recognize human speech, and can speak back. Notably, when humans know they are interacting with a chatbot, their behavior changes to be more rude, less open, and to using shorter phrases.[2,3]

Digital Innovations in Healthcare Education and Training.
http://dx.doi.org/10.1016/B978-0-12-813144-2.00008-8
Copyright © 2021 Elsevier Inc. All rights reserved.

Chatbots for educational purposes
Chatbots for impacting learning outcomes

Multiple applications for chatbots have been considered in Education, which include: quizzing existing knowledge,[4] asking basic factual knowledge in specific topic areas[5] for creating higher student engagement with a learning task,[6] and for mentoring success among first year university students.[7] These chatbot examples in higher education focus on factual information drawn from a database for a knowledge-level learning goal[8] (Fig. 8.1. Bloom's Taxonomy with examples of AI for higher education). That is, knowledge-level chatbots provide a similar learning experience to a textbook or website, where learners may seek out answers to questions about facts and information. Such chatbots do not, however, help learners complete advanced tasks typically expected within university programs.

Bloom's Taxonomy[8] outlines cognitive processes in a continuum from basic to more advanced. Higher-order cognitive activities, such as *apply, analyze,* and *evaluate* are necessary and important goals of education. Can chatbots be used to assist more advanced learning? Multiple projects reported teaching learners a new language through chatbots.[9,10] Other projects consider meta-cognition prompts for learners to better understand their own learning habits[11] or prompt learners to reflect on their course work.[12] Such chatbot projects not only support higher-order cognitive activities, but give the sense that chatbots might be able to support learner success through simple question–and–answer means.

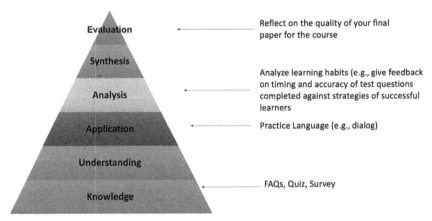

Figure 8.1 *Examples of AI for higher education and where align on Bloom's Taxonomy.*

Chatbots for efficiently supporting learners and scaling faculty

Chatbots provide a distinct advantage over other digital formats for learner support. Basic digital support for higher education learners comes in the form of websites with detailed tips on performing well in their courses, or where to access help; alternatively, live office hours through web conferencing with faculty and/or support staff might be used to help learners to success. In contrast, chatbots provide a direct response to learners, in real time, and at any time of day—that is, even at night, when the teacher or faculty course leader is reasonably inaccessible. Such tools have been considered to support learners at a distance, including eLearning courses.[13,14] Health professions programs bring the "perfect storm" of needs for chatbot applications: busy learners distributed across clinical and university sites, who have needs to connect back to campus resources to support their success. Given required service on evenings and nights, rural or urban distributed clinical sites, and the time-pressed nature of health professions education, learners cannot always connect with faculty or staff during normal business hours, and/or have very limited time in their days in which to search the web for answers. Chatbots thus provide a solution to support health professions learners 24×7 and across geographically distributed clinical and institutional sites.

Similarly, chatbots provide an advantage to scaling faculty. In an era of faculty burnout,[15,16] scaling faculty that they may have time and bandwidth to focus on those tasks that energize them and allow them to succeed among multiple pressure to produce is essential.[17,18] One area of scaling faculty is to focus on those tasks that could be done by other chatbots. Repetitive tasks contribute to the type of boredom that negatively impacts performance.[19] There are many instances of repetition in higher education teaching, including encountering repetitive learner questions. It is particularly the example of repetitive learner questions that could be done well by the chatbot that the faculty member may refocus on more engaging parts of their role. Second, health professions education programs are programs known for delivering dense amounts of content with high expectation of recall. To effectively achieve learning outcomes in such a time-pressed, content-dense learning environment, there is a fundamental change required in learning design of the program, where faculty and learners engage in active, higher-order cognitive processes.[20] This too indicates that Bloom's Taxonomy levels of remembering and understanding could be conducted through an autonomous agent, such as a chatbot. We therefore focus our attention on what tasks might be completed by chatbots in health professions education programs in order to support learners and scale faculty.

Chatbots for health professions education

Combining both the continuum of cognitive processes presented in Bloom's taxonomy[8] with the idea that some repetitive tasks in health professions education programs might be done with a chatbot to provide greater access or to scale faculty time, we present a table of opportunities for chatbots (Table 8.1). Basic health sciences factual recall chatbots are presented at the bottom of the table, reflecting the information-level exchange of

Table 8.1 Possibilities for chatbots in health professions education.

Bloom's Taxonomy	Examples with description
Creating	*Practicing Night Call.* Through the chatbot (which is the attending on call), conduct an over-the-phone consult to practice both one's own communication skills, as well as evaluate questions coming from the "attending" (the chatbot), while synthesizing and replying in a timely manner based on patient information known to the learner and provided at the start of the Night Call experience. *Practicing an IPE consult.* Trainees of various health professions backgrounds are able to simultaneously or independently explore a patient case from multiple health care perspectives in order to gain empathy and professional understanding of the other professions.
Evaluating	*Practicing a patient consult.* A full patient case can be explored through dialog with the chatbot posing as the patient.
Analyzing	*Practicing lab interpretation.* Given physiological data, learners are able to practice assessing what is happening with the patient.
Applying	*Advanced anatomy chatbot,* where learners may ask about connections between anatomical structures or functions of anatomical structures. For example, the Voxscholar Anatomy chatbot (voxscholar.com).
Understanding	*Study Skills chatbot,* where learners may share about their exam performance difficulties to get suggestions about how to improve. Learners may also ask about evidence-based strategies for studying that improve memory and understanding of the content taught. For example, the Voxscholar Study Skills chatbot (voxscholar.com). *Faculty Development chatbot.* This chatbot uses the faculty development evidence base to provide short prompts to faculty to coach them to successful performance as educators.
Remembering	*Anatomy chatbot,* where learners may ask basic questions about muscles, nerves, vessels, or bones. For example, the Voxscholar Anatomy chatbot (voxscholar.com).

remembering content. As one reads through to the top of the table, higher-order cognitive processes are required through interaction with the chatbot.

Focusing on the remembering and understanding rows near the bottom of Table 8.1, the chatbots for study skills and anatomy are anticipated to scale faculty time by answering questions in place of the instructor. For example, in anatomy lab, multiple students may have questions at the same time, which is difficult for the same instructor to physically attend to all learner dissection tables at the same time. Thus, the anatomy chatbot is able to answer most commonly asked questions, so learners call on faculty when there are more complex questions, or questions specific to the cadaver patient, the trainees are working with. Learners may also access such a chatbot before and after anatomy lab times, which can support their learning. Similar efficiencies for faculty and trainees are anticipated for the example of the advanced anatomy chatbot in the *applying* row of Table 8.1.

Similarly, the study skills chatbot answers learner inquiries about their typical performance issues on exams (e.g., running out of time, having difficulty understanding multiple choice questions) that are repetitively answered by faculty or learning success staff in student support services. Learners might also improve their studying skills by learning evidence-based tips from the chatbot. Both functions allow learners to independently seek help in frequently encountered performance issues, so faculty and/or learning success staff are able to focus on more advanced conversations with learners in difficulty. Such chatbots support efficient encounters, which may lead to cost-savings for the institution, while also providing learners with access to supportive resources, no matter the time of day (or night)—or proximity to an exam. These chatbots might also reduce student stress and improve trainee wellbeing by being accessible at the point of need.

A third example, the faculty development chatbot, reduces faculty burnout by providing just-in-time, evidence-based answers on topics related to faculty development. The multiple demands placed on faculty at academic health centers often mean that faculty have to make choices between attending multi-hour faculty development workshops, and attending to patients or teaching trainees. Most evidence-based teaching practices can be effectively offered in short bursts, such as bulleted lists of tips, or short videos (1–5 minutes) demonstrating the skills involved.[21] As chatbots are able to deliver both text-based tips and videos to mobile devices, faculty may ask questions right before, or in the moment of teaching. For example, the busy preceptor who arrives at the clinic and remembers they teach a student today is able to quickly ask for tips on specific topics (e.g., how to set expectations, how to give effective feedback) prior to meeting with the trainee.

Notably some of the examples in Table 9.1 might be more efficient for faculty time *and* provide learners with more in-depth deliberate practice experiences of important features of clinical encounters. Four of the examples suggested in the *analyzing, evaluating,* and *creating* rows of Table 9.1 relate to the practicing elements of patient care. Trainees are often limited by the scope of patients present in clinic, with their learning further compounded by the rapid turnaround expected within patient interviews (often 5–15 minutes per patient) within clinical settings. This uncontrolled content of patient cases and time-pressed nature of learning in clinical settings is opposite of deliberate repetition of integrating skills and knowledge in specific domains toward mastery, also known as deliberate practice.[22] Virtual patients have used in multiple health professions contexts to provide deliberate practice toward improving trainee clinical reasoning.[23,24] Chatbots offer the opportunity to advance virtual patients to a true conversation, rather than screen-based text-to-computer communication formats. As chatbots may be deployed to mobile devices and speech devices like Alexa (Amazon, Seattle, Washington, USA), Google Home (Google.com, Cupertino, California, USA), and Siri (Apple, Cupertino, California, USA), learners may participate in deliberate practice of their skills in formal learning settings, as well as on their own time. As health professions education focuses on competent performance,[25] chatbot-enabled virtual patients will be increasingly important to develop learner competencies.

Artificially intelligent chatbots for HPE: the VoxScholar project

The basic chatbot question-and-answer format does not suffice for all of the applications in Table 9.1. Virtual patients in chatbot form require artificial intelligence (AI) as well as machine learning necessary to deliver responses more reflective of typical human-to-human conversations. AI refers to the development and use of computer systems to complete tasks that normally require human intelligence. Speech recognition, which is a feature of the VoxScholar chatbots, is one type of AI. AI is also known to be involved in decision making, which can be as simple as determining which answer to deliver to a chatbot user next, as in the case of answering a student's question about study skills. More advanced decision making AI might involve analyzing past behavior of a chatbot user, and delivering more personalized content in subsequent interactions. The faculty development chatbot is designed on such a platform, where the individual faculty member's usage of the chatbot is tracked over time to deliver more advanced content over time. The chatbot and AI platforms analyze multiple points of data (e.g., time of day, topic asked, user satisfaction rating at end of the chatbot conversation) in order to determine what depth of content a faculty member might need next. For example, if the faculty member asked about giving effective feedback 6 weeks ago when they had a learner in a community-based clinic, when the faculty member reconnects with the chatbot when a new trainee joins the clinic, the chatbot will prompt with the past tip and more advanced tips for the faculty member to select from. Based on the faculty member's selection, the AI algorithm will make adjustments and calculations for the next interaction. Over time and with increasing numbers of users, the AI system improves in its "smart delivery" of content. The ideal would be that the AI algorithm starts learning which faculty needs more advanced tips over others, but always moving every *individual* along a path to expertise and mastery over time

Lessons learned with chatbots in health professions education

There is much to be evaluated with chatbots in health professions education, particularly chatbots powered by AI. The VoxScholar project is currently evaluating multiple chatbots with trainees and faculty. Early lessons learned have been that while excitement to use chatbots is high, not all users engage with the chatbot equally, and prompts are needed to encourage engagement over time. Learners are particularly excited to use a chatbot rather than consult a human faculty or staff member, a finding that is supported by the literature that millennial learners prefer text-based communication for talking to humans. Security requirements of university and clinical site wireless networks require special customized networks if voice-enabled hardware like Amazon Echo (Amazon, Seattle, Washington, USA) or Google Home (Google, Cupertino, California, USA) are used to offer the voice-based chatbots. Moving to voice-enabled apps loaded to users' mobile devices, such as the Alexa App (Amazon, Seattle, Washington, USA) or the Google Assistant App (Google, Cupertino, California, USA), provides access while also removing costly institutional investments in hardware as well as wireless infrastructure. The costs, however, may be passed on to users through their mobile device plans.

Multiple levels of evaluation are planned for the VoxScholar project. Content-specific chatbots, such as the anatomy chatbot, will be evaluated for learning outcomes in the domains of: *knowledge, understanding*, and *applying*. Task support chatbots, such as study skills and faculty development, will look at impact on behavior over time. All chatbots will be evaluated for user-centered design, and continuous quality improvement is in place to improve conversations.

Conclusion

This chapter has outlined how chatbots provide promising opportunities to scale faculty and trainee time and learning through just-in-time access to content and answers. Chatbots might be also be an accessible option for reducing faculty burnout, improving trainee access to central resources while in geographically distributed clinics, and engaging in deliberate practice. While many applications of chatbots are possible, programs are encouraged to focus on those that have direct impact on learning outcomes. Using Bloom's taxonomy as a guide has proven helpful in this regard.

It is anticipated as chatbots gain traction in home-based hardware devices provided by major software companies like Apple, Google, and Amazon, that our learners and faculty will have functional expectations for higher education programs to adopt chatbots as well. The core challenge ahead of all of our programs is to purposefully develop educationally impactful chatbots that advance evidence-based behaviors, while not adding to stress and burnout among our most valued assets: our faculty and trainees.

References

1. Shawar BA, Atwell ES. Using corpora in machine-learning chatbot systems. *Int J Corpus Linguist* 2005;**10**(4):489–516 doi:https://doi.org/10.1075/ijcl.10.4.06sha.
2. Hill J, Ford WR, Farreras IG. Real conversations with artificial intelligence: a comparison between human–human online conversations and human–chatbot conversations. *Comput Hum Behav* 2015;**49**:245–50.
3. Mou Y, Xu K. The media inequality: comparing the initial human-human and human-AI social interactions. *Comput Hum Behav* 2017;**72**:432–40.
4. Pereira J. Leveraging chatbots to improve self-guided learning through conversational quizzes. *Paper presented at the Proc. of the fourth international conference on technological ecosystems for enhancing multiculturality*; 2016.
5. Kazi H, Chowdhry B, Memon Z. MedChatBot: An UMLS based chatbot for medical students. *Int J Comput Appl* 2012;**55**(17).
6. Heller B, Proctor M, Mah D, Jewell L, Cheung B. Freudbot: An investigation of chatbot technology in distance education. *Paper presented at the EdMedia: world conference on educational media and technology*; 2005.
7. Augusto JC, McNair V, McCullagh P, McRoberts A. Scoping the potential for anytime-anywhere support through virtual mentors. *Innovat Teach Learn Informat Comput Sci* 2010;**9**(2):1–12.

8. Anderson LW, Krathwohl DR, Airiasian W, Cruikshank K, Mayer R, Pintrich P. *A taxonomy for learning, teaching and assessing: a revision of Bloom's Taxonomy of educational outcomes: complete edition.* NY: Longman; 2001.

9. Jia J. CSIEC: a computer assisted English learning chatbot based on textual knowledge and reasoning. *Knowledge-Based Syst* 2009;**22**(4):249–55.

10. Zakos J, Capper L. *CLIVE–an artificially intelligent chat robot for conversational language practice.* Berlin, Heidelberg: Springer-Verlag; 2008.

11. Kerly A, Ellis R, Bull S. CALMsystem: a conversational agent for learner modelling. *Applications and innovations in intelligent systems XV* Springer, London: Springe; 2008. p. 89–102.

12. Grigoriadou M, Tsaganou G, Cavoura T. Dialogue-based reflective system for historical text comprehension. *Paper presented at the workshop on learner modelling for reflection at artificial intelligence in education;* 2003.

13. Bollweg L, Kurzke M, Shahriar KA, Weber P. *When robots talk-improving the scalability of practical assignments in MOOCs using chatbots.* Paper presented at the EdMedia+ Innovate Learning.

14. Lundqvist KO, Pursey G, Williams S. Design and implementation of conversational agents for harvesting feedback in eLearning systems. *Paper presented at the European conference on technology enhanced learning;* 2013.

15. Lackritz JR. Exploring burnout among university faculty: incidence, performance, and demographic issues. *Teach Teacher Educ* 2004;**20**(7):713–29.

16. Pololi LH, Evans AT, Civian JT, Gibbs BK, Coplit LD, Gillum LH, et al. Faculty vitality—surviving the challenges facing academic health centers: a national survey of medical faculty. *Acad Med* 2015;**90**(7):930–6.

17. Shah DT, Williams VN, Thorndyke LE, Marsh EE, Sonnino RE, Block SM, et al. Restoring faculty vitality in academic medicine when burnout threatens. *Acad Med* 2018;**93**(7) doi:10.1097/acm.0000000000002013.

18. West CP, Dyrbye LN, Erwin PJ, Shanafelt TD. Interventions to prevent and reduce physician burnout: a systematic review and meta-analysis. *Lancet* 2016;**388**(10057):2272–81.

19. Haager JS, Kuhbandner C, Pekrun R. To be bored or not to be bored—how task-related boredom influences creative performance. *J Creat Behav* 2016;**52**(4):297–304.

20. McNamara DS. Strategies to read and learn: overcoming learning by consumption. *Med Educ* 2010;**44**(4):340–6.

21. Corral J, Post MD, Bradford A. Just-in-time faculty development for pathology small groups. *Med Sci Educ* 2018;**28**(1):11–2.

22. Ericsson KA. Acquisition and maintenance of medical expertise: a perspective from the expert-performance approach with deliberate practice. *Acad Med* 2015;**90**(11):1471–86.

23. Cook DA, Erwin PJ, Triola M. Computerized virtual patients in health professions education: a systematic review and meta-analysis. *Acad Med* 2010;**85**(10):1589–602.

24. Ellaway RH, Davies D. Design for learning: deconstructing virtual patient activities. *Med Teacher* 2011;**33**(4):303–10.

25. Frank JR, Snell LS, Cate OT, Holmboe ES, Carraccio C, Swing SR, et al. Competency-based medical education: theory to practice. *Med Teacher* 2010;**32**(8):638–45.

CHAPTER NINE

Learning analytics, education data mining, and personalization in health professions education

Janet Corral[a], Stathis Th. Konstantinidis[b], Panagiotis D. Bamidis[c]
[a]University of Arizona Tucson College of Medicine, Tucson, AZ, United States
[b]University of Nottingham, Nottingham, United Kingdom
[c]Aristotle University of Thessaloniki, Greece

Chapter outline

Personalization of health professions education through learning analytics and education data mining

This chapter is meant as an introduction to personalization of health professions education (HPE) through educational analytics. To start, the chapter will explore the definitions of learning analytics and education data mining in terms of their approaches and relationships with student data. However, most academic health education programs have challenges in collecting, storing, and analyzing education data. These issues will be addressed before exploring the various ways in which analytics may be personalized to the individual learner, including how these forms of individual analytics of each learner may be displayed to administrators or course leaders to

Digital Innovations in Healthcare Education and Training.
http://dx.doi.org/10.1016/B978-0-12-813144-2.00009-X
Copyright © 2021 Elsevier Inc. All rights reserved.

inform decisions about the education content, teaching, or future educational design.

Introduction to analytics in health professions education

Learning analytics is the "measurement, collection, analysis, and reporting of data about learners and their contexts, for purposes of understanding and optimizing learning and the environments in which it occurs."[1] This approach to education data analysis is keenly focused on understanding the context, environment, and social connections in the educational experience that are also influencing learning.[2] Learning analytics, as a practice, emphasizes a holistic approach to data analysis, which involves applying human judgment and intervention both at the outset, before data is collected, as well in the analysis, to better understand learning systems as whole, complex systems involving learners and contexts. Learning analytics takes the perspective that the data on any individual or group of learners is not solely about their learning outcomes (e.g., exam scores or progress), but is fundamentally embedded in the educational environment replete with significant others (e.g., other learners, teachers), tools (e.g., books, calculators, stethoscopes, etc.), and contexts (e.g., lecture theatre, patient bedside).

A similar and related discipline, education data mining (EDM), also focuses on learning, but in a different way. EDM is defined as "exploring the unique types of data that come from educational settings, and using those methods to better understand students, and the settings which they learn in."[3] EDM takes a reductionist, automated approach to analyzing data learners generate within computer-based systems for learning, such as drill-and-practice math problems or standardized exams. Published examples include: series of learner answers on radiology image readings that are automatically analyzed into learning curves,[4,5] or predicting the selection of science as a major in university based on elementary school mathematics,[6] or predicting which students will drop out of school by the end of elementary school.[7] EDM has a greater focus on modeling student progress, predicting student progress, and on creating computer systems that adapt without a human needing to intervene in the learning cycle (e.g., intelligent tutors). EDM approaches may provide the analyzed data back to learners, such as through an intelligent tutor system to change and personalize a learner's experience.[3] In contrast, the learning analytics approach to similar data would

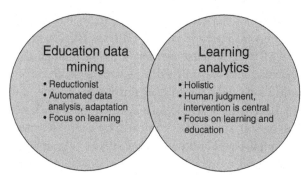

Figure 9.1 Defining and distinguishing between learning analytics and education data mining. *Corral J, 2015 based on Ref. [3].*

more closely tie to theory and try to understand the relationship among the constructs in the learning system. Learning analytics would also try to inform and empower instructors and learners, for example: informing instructors about ways that specific students are struggling, so that the instructor can contact the learner[3] (Fig. 9.1).

Does HPE have the data to do learning analytics and EDM?

Big data is defined by four variables: volume, veracity, variety, and velocity.[8,9] Big data volume is measured in zetabytes (i.e., trillion gigabytes); velocity relates to the speed of analyzing the data; variety is defined by different forms of data; and veracity is the accuracy of the data. Learning analytics and EDM are often mistaken as methods for "big data in education"; yet, with sample sizes and speeds of learner data not nearly close to the definition of "big data,"[8,9] the caution we put forth in HPE learning analytics is to appreciate our relatively humble—but able—scale.

In order to fully appreciate analytics in HPE, it is also important to spend time on understanding the context of HPE and its data. HPE programs have very complex learning contexts. HPE programs span both the higher education institutions (e.g., universities) and the clinical practice milieus (e.g., hospitals, clinics, telehealth). As we will explore later, this has important implications for the speed and veracity of the data collected. Given the international audience for this chapter, we will discuss only the generic contextual issues (i.e., general interacting components and actors), rather than specifics (e.g., the duration of a program, the multiple possible entry and exit points of learners).

The complexity of the HPE learning contexts has implications to the foundations of personalized analytics in HPE in four key ways:

1. *Data is spread across multiple systems.* The technological infrastructure spans university, hospital, and national databases. University and third-party applications paid for by the university are the most accessible sources of data (e.g., learning management system, learning content system, assessment system for clinical rotations, assessment system for exams). Hospital-based and national databases (e.g., graduate questionnaire, standardized exam results, residency matching data, certification exam results) may or may not be accessible to the HPE for inclusion at the data warehouse level, which is essential to being able to analyze for personalization. If access is possible, access to these latter sources comes following long, careful negotiations between institutions.

2. *Data veracity is complicated.* The various databases storing the various types of evaluation data may be "out of synch" in terms of when the data was collected. A complicating factor in clinical rotations is that individual assessments of student performance in clinic are returned over a time period of weeks, rather than by the assigned deadline. Fundamental to evaluation in HPE, whether for analytics or not, is that not all evaluators apply the assessment tools in aligned with program requirements.[10] Each of these factors has implications for the veracity of the HPE data in HPE data warehouses.

3. *Speed of data reporting is impacted by data veracity, and IT infrastructure.* Even rapid teams in HPE anecdotally report taking half a year to a year to enable and populate their data warehouses with existing institutional data. Standards for data across the contexts of HPE are currently in process through the MedBiquitous Consortium[11] and IMS Global.[12] These standards will make it easier to analyze across curricula,[13] at the level of the learner or learning taking place.

4. *Data volume is low within individual HPE programs.* Since class sizes are limited to a few hundred students in medical school, and for postgraduate programs, less than a hundred participants per specialty, most data for medical education fails to approach the millions or billions of data points seen in true Big Data sets. Longitudinally, a medical school or residency program might have data points in the thousands or tens of thousands—though veracity may be impacted as curricula change. The same data size concern persists for most higher education data, though larger data sets are possible with online and mobile learning in massively open online courses from EdX, Coursera,[14] or from book publishers of standard

elementary, high school, and general higher education textbooks,[15,16] as well as within dedicated storehouses collecting learner data, such as the Pittsburgh Science of Learning Center DataShop.[17] Within HPE, organizations running credentialing exams,[18] or multi-institutional learner data, like MedU (https://www.med-u.org/) or Entrada (www.entrada. org) will have more robust data sets for analysis.

These four considerations culminated to another important point: HPE data has a certain level of "messiness." This "messiness" may slow the proficiency of IT teams in HPE, and thus the analytics possible. However, as we will see through the examples explored in the next section, it is possible to work with the *best available data* to implement personalization projects in HPE.

Personalization, feedback, decision-support, and self-directed learners

Personalization is customization to the individual learner. Such customization may involve providing feedback on performance; coaching performance through feedback; providing future questions based on past performance; or predicting future performance relative self, or a larger group of learners. Nascent and critical to the discussion of personalization is to acknowledge personalization requires sharing the data analysis results *with the learner*. Personalization is not personalization unless the learner is involved.

Feedback

When learner data is analyzed and displayed back to the learner as a visual dashboard of grades or student progress (in a course, clinical rotation, assignment, etc.), personalization becomes a form of feedback. This approach provides both a level of transparency to the learner, as well as access to longitudinal progress to date. Notably in HPE, new data comes often at a time delay of days (e.g., exam results or OSCE results), to weeks (e.g., clinical rotation results). Thus, these forms of learner progress may be more like successive iterations of punctuated feedback to learners, rather than live-streamed progress updates.

Spickard et al.[19] provide an example of implemented visual dashboards of learner grades as they progress through a medical school program. Vanderbilt Medical School has implemented a competency-based approach to assessing learners, supported by a new cadre of faculty called "coaches." The coaches can see the individual progress of learners assigned to them; learners can only see their personal progress; administration can see individual learners, the whole class, and across competencies of the medical school program. This type of implementation exemplifies the multiple roles needed to successfully implement personalization as feedback. This implementation, while a laborious undertaking to build and execute, reflects the simplest of the examples of personalization: visual displays of assessment data, refreshed only when there are new data to display.

One consequence of providing dashboards to learners, faculty, and administration is that learners need to interpret, critically appraise, and act on the feedback they receive.[20,21] Visual literacy (interpretation of visualized dashboards) is essential in order for learners, their coaches, and their administrators decision makers to support learning. Support within the HPE program for the continuum of adopting dashboards for learner feedback is necessary. This includes: orientation of learners to receiving, appraising, and acting on feedback; coaching support during the HPE program to ensure learners are appraising the feedback appropriately, and knowing the wealth of resources to support appropriate actions; and lastly, support for faculty coaches to learn the coaching skills necessary will be key to ensuring student success. Simultaneously, it is important for administrators to make appropriate decisions around privacy of the learner data, particularly when it impacts future career match by sharing the dashboards (or the data from the dashboards) across to future employers (e.g., between undergraduate and residency programs in the United States and Canada; between hospital and healthcare systems in the European Union and Australia).

Predicting learner success

When past data predict future learner progress, or even success, that is fed back to the learner and/or administration, personalized prediction is the result. Personalized prediction is a sub-field of analytics that uses one of several methods (e.g., linear regression, logistic regression, Bayesian networks) with existing data, to predict future outcomes.[22] This iterative process may use longitudinal data (e.g., years of regional standardized exam scores[6]) or immediate data (e.g., student activity on day 1 of an online course to predict success and course completion[23]). The prediction process requires a training set (e.g., data from past learners), from which an algorithm is developed and tested on a new data set, called the testing set (i.e., data generated from current learners whose data was not in the training set). To successfully predict future learners' success, the training set must be sufficiently large; from learners very much like future learners; have quantifiable characteristics to statistically model; and a clear outcome of interest.[22] Whether a predictive model is successful or not depends on its predictive accuracy, that is, success at predicting student response outcomes, predicting post-test outcomes, or pre-test/post-test gains.[24]

Prediction is hard work. In the example of predicting how many electrocardiograms residents need to read in order to be masterful, there are years of work and several iterations of interventions with learners, to gather sufficient data to create learning curves[4,5]—which are the training set, not the predictive model. The learning approach may still result in unclear prediction models for two reasons. First, there are multiple confounding variables in determining why a learner read an electrocardiogram incorrectly (e.g., was it a guess, was it inability to distinguish the angle of the trace correctly, was the trace more difficult than the learner could handle?). Second, in learning curve work in general higher education with much larger data sets, researchers are still puzzled as to what learning curves actually mean in terms of learner engagement and learning outcomes.[25,26]

Despite these challenges, there are great opportunities for prediction in HPE. Particularly for the high-stakes standardized exams and matching to future training programs, prediction may leverage larger data sets over multiple years of learners to produce more reliable outcomes. Within single HPE programs, online quiz and exams data provide the ability to predict student performance on a more granular scale, so that learners get early warning about topics they need to study more closely, or competencies that need more attention.

Adaptive personalization

Adaptive learning holds the promise that instruction or examination can be personalized to meet the learner's level of difficulty. For example, the licensing exam from the Medical College of Canada provides questions of different difficulty based on how examinees respond to earlier question sets (http://mcc.ca/examinations/mccee/). Similarly, several third-party applications provide sets of practice questions, accompanied by feedback, to help medical students strategically prepare in content areas, they need the most practice in for the USMLE Step 1 exam (e.g., Firecracker; http://www.firecracker.me/). Adaptive learning systems for HPE content are also beginning to emerge (e.g., Macrophage.co; http://www.macrophage.co/).

Developing adaptive systems requires both the analysis of learners' progress (whether from past data, or present data created in the system), combined with expert insights about rating content difficulty (e.g., question psychometrics, difficulty of scientific content) and determining success (e.g., cut-off scores to progress to the next "level"). The systems also need to be tested and validated. This can be a lengthy process, particularly when, within HPE, the expertise of many content areas (e.g., interpreting radiology images) are based on learning as an apprentice from a more knowledgeable other. As such, much lead-in work needs to be done regarding the difficulty and effort of the learning content, task, and categorizing the correct answer,[4,5] as well as developing the coaching feedback to embed into the final intelligent tutoring system. While this may appear daunting, a competing direction is the advanced nature of artificial intelligence in certain clinical disciplines, such as oncology[27] and primary care,[28] where the computer (or, rather high performance computing and/or machine learning) is able to successfully interpret the patient history and investigation results, and provides a recommended diagnosis or next step to the MD. Such simultaneous advancement challenges the types of intelligent tutors, as well as content mastery, our HPE programs should be building.

Adaptive learning systems do raise awareness among educators of the question: "what matters?." Traditional measures of learning outcomes focus on achieving a passing grade. Intelligent tutors provide insightful measures of engagement, individual agency, and time on task, which have been found to be more highly predictive of progress or success in online education.[29-31] Designing such systems to incorporate a fuller feedback loop between educator and learner requires both collecting affective information and incorporating prompts to instructors when learners are showing frustration, engagement or concentration.

Structuring data-informed conversations and action in health professions education

Providing personalized feedback—whether as a dashboard, a predictive algorithm estimate, or adaptive system response—provides "new" information (or, old information in a more timely or transparent manner) to learners, faculty, and administration. This will undoubtedly perturb the preexisting ways in which student evaluation was disseminated, as well as the roles of faculty and administration, who may now be expected by learners to be responsive in ways that were not previously asked of their roles. This was true at Purdue University, when the institution introduced the Course Signals system.[32] Course Signals which was a simple dashboard reporting green (ok), yellow (caution), and red (negative alert) "street signals" to learners and faculty about learner's progress in specific undergraduate courses. This approach led to learners who received red signals immediately contacting professors about their grades in a higher volume than previously; Purdue hired a larger number of student advisors to provide assistance to learners.[32] What impacts might personalization have on health professions education faculty and curriculum operations?

Vanderbilt University School of Medicine recently adopted a new curriculum concurrently with personalized analytics of learners.[33] A whole new category of faculty, called "coaches," whose main role is to support meta-reflection and provide guidance to learners, was concurrently created with the personalized dashboards. Such a synergistic evolution is perhaps more novel among the medical schools than typical; many HPE programs are playing "catch-up" as they adopt personalization efforts with existing curricula and faculty roles.

So how might HPE programs most effectively engage in data-informed conversations about, for and with learners based on personalization?

1. **Develop a data-informed culture for making instructional decisions.** Many curricular and assessment decisions at academic health centers are discussed in terms of the existing literature, and internal learner, faculty, or program evaluation data. This is a great starting point to discuss the roles and the "how" of the committees, faculty, learners, administrators, and staff will engage with personalized analytics.

2. **Determine which analytics are most impactful to your program.** Just as our academic medical community debates which learning outcomes should be reported in research studies, so too should or analytics programs focus on metrics that are meaningful to stakeholders— including learners. Our community also needs to consider analytics data to *improve learning* with a different mindset engendered by data systems created for *accountability* purposes.

3. **Help stakeholders understand the source and usefulness of personalized learning data.** Having discussions about where the data came from, how the data is used by the learning system, and how stakeholders should/could use the data to inform their decisions (e.g., what to study, changes to instruction, changes to assessment) is paramount. Walking stakeholders through the visual dashboards will help with definition setting and setting a common understanding of actions to take based on the data being reported.

4. **Recognize that analytics is a program that needs adequate resources.** The process of building a data warehouse, curating the data,

conducting the analysis, and reporting the results to stakeholders with differing levels of detail is a time-consuming undertaking that also requires investment in people and computing resources. Even when leveraging open source options, there are costs to hosting the data and developing expertise in HPE personnel. Each report, each algorithm, will take time and resources. Making analytics a cost-effective undertaking requires a clear decision-making structure that routinely revisits which metrics and parties need the analytics being requested, and to what benefit to the organization.

5. **The academic or information technology team is part of the effort, but not the only responsible department.** Assessment, curriculum, instructional staff, student representatives, and top decision makers need to work collaboratively on owning personalized analytics initiatives. Personalization analytics are an intervention in the educational program that will only reinforce, rather than create separate from, the intricate relationships between learners, instructors, assessment, and educational leadership.

Limitations and concerns for personalization within HPE

Education data mining of online learners' data has been criticized as seeing learners as "disembodied autodidacts."[34–36] The reductionist nature of data-based representation is a core limitation of HPE personalization, and should be particularly heeded at time when much of the HPE literature is focused on competencies in non–cognitive attributes, such as empathy, collaboration, and communication skills.[37,38] More bleeding edge work in education data mining tries to marry qualitative observations of learners with their computer or web log data to make determinations about how the observed learner emotions relate to their data points and progress within computer- or web-based training systems.[39] This work acknowledges emotion is part of learning, and aims to determine when frustration, engagement, and boredom can be detected and reflected back to the learner to support the self-regulation of learning.[40,41] Early work with middle school learners' engagement with online math software has been predictive of later college enrollment.[6,42] This work has not yet been applied to HPE contexts, providing a gap for future research that should be carefully explored given the sensitive nature of student data.

This brings us the concerns about how student data is obtained in HPE. Generally in higher education, there are concerns related to the privacy and ethics of using, analyzing, and visualizing student data, given that the learner is essentially forced to provide their data as they use learning systems required by the institution.[43–45] These concerns persist after the learner has left the institution, as the data is held in data warehouses to allow for further predictive modeling for personalizing the learning experience for future cohorts. Notably, HPE is a high-stakes field, particularly at the transition between early training (e.g., in undergraduate HPE) and clinical immersive training (e.g., postgraduate or residency training). Should HPE programs retain the data for continuous quality improvement of the program? For honing algorithms supporting personalization—and possibly better learning outcomes? These are questions that will be answered differently based on cultural and legal contexts of different jurisdictions.

Future considerations for personalization in HPE

While general higher education programs have been working in EDM and learning analytics toward personalization since the early 2010s, the first analytics sessions at national conferences in HPE were in 2014; the first conference dedicated to data-driven academic health centers was in 2015. The HPE community is perhaps late to the table, and hindered by small warehouses of education data drawing from multiple sources and contexts across clinical and traditional university learning settings. Collectively, those interested in HPE need to ask: which types of personalization make sense for HPE to engage with? Should we start where higher education started, or should we use the more learning analytics-friendly approach of understanding our context, and "jumping in" with the more bleeding edge work?

Newer models of personalization appreciate the fundamentally interconnected nature of learning. That is, the social contexts, whether learning from more knowledgeable others in informal settings or more formalized (and imposed) contexts like small group learning or group projects, impact the individual. Connecting performance of the individual learner to both the pedagogies used, as well as the social structure within a course, is showing promise of capturing this dynamic interplay of knowledge building.[46] This provides a key gap for HPE analytics practitioners to consider adopting.

Conclusion

The field of personalization in HPE programs is relatively new. Learning analytics, which focuses on understanding the context of learning, and education data mining, which implores a more automated-analysis-first approach, both have contributions to the personalization of HPE. This chapter has presented how personalization may manifest as feedback, predicting learner success, or adaptive learning, notwithstanding the fundamental challenge that HPE data is not Big Data, and so many more realistically relay on traditional statistical methods. A key issue addressed in this chapter is acknowledging that personalization, in any form, will likely perturb preexisting roles and operations within HPE programs; leaders should be cognizant and plan for how they will lead change. The field of personalization is ripe with possibilities for HPE; the closing question to all practitioners, leaders, and learners remains: which approaches benefit our community most, and toward what end?

References

1. Siemens G, Long P. Penetrating the fog: analytics in learning and education. *EDUCAUSE Rev* 2011;**46**(5):30.
2. Baker RS, Inventado PS. *Educational data mining and learning analytics. Learning analytics.* New York, NY: Springer; 2014.
3. Siemens G, Baker RS. Learning analytics and educational data mining: towards communication and collaboration. In: *Proceedings of the second international conference on learning analytics and knowledge*; 2012.
4. Pusic MV, Boutis K, Hatala R, Cook DA. Learning curves in health professions education. *Acad Med* 2015;**90**(8):1034–42.
5. Pusic M, Pecaric M, Boutis K. How much practice is enough? Using learning curves to assess the deliberate practice of radiograph interpretation. *Acad Med* 2011;**86**(6):731–6.
6. Pedro MO, Ocumpaugh J, Baker R, Heffernan N. Predicting STEM and non-STEM college major enrollment from middle school interaction with mathematics educational software. In: *Educational Data Mining*; 2014.
7. Knowles JE. Of needles and haystacks: building an accurate statewide dropout early warning system in Wisconsin. *J Edu Data Mining* 2015;**7**(3):18–67.
8. De Mauro A, Greco M, Grimaldi M. A formal definition of Big Data based on its essential features. *Library Rev* 2016;**65**(3):122–35.
9. IBM. What is Big Data? *Big Data.* Available from: https://www.ibm.com/big-data/us/en/; 2016 [Accessed March 27, 2017].
10. Konstantinidis ST, Bamidis PD. Why decision support systems are important for medical education. *Healthc Technol Lett* 2016;**3**(1):56–60.
11. Consortium M. *Collaborative technologies for medical education.* Baltimore, MD: The MedBiquitous Consortium; 2003.
12. Abel R, Kellen V. *Simplifying learning analytics via the caliper analytics framework*; 2015.
13. Ellaway RH, Albright S, Smothers V, Cameron T, Willett T. Curriculum inventory: modeling, sharing and comparing medical education programs. *Med Teacher* 2014;**36**(3):208–15.
14. Jordan K. Initial trends in enrolment and completion of massive open online courses. *Int Rev Res Open Distribut Learn* 2014;**15**(1).
15. DiCerbo KE, Behrens JT, Barber M. *Impacts of the digital ocean on education.* London: Pearson. Retrieved September. 2014; 1:70–81.
16. DiCerbo KE. Game-based assessment of persistence. *J Edu Technol Soc* 2014;**17**(1):17–28.
17. Koedinger KR, Baker RS, Cunningham K, Skogsholm A, Leber B, Stamper J. A data repository for the EDM community: the PSLC DataShop. *Handbook of educational data mining* Boca Raton: CRC Press; 2010. p. 43.

18. Ellaway RH, Pusic MV, Galbraith RM, Cameron T. Developing the role of big data and analytics in health professional education. *Med Teacher* 2014;**36**(3):216–22.
19. Spickard III A, Ahmed T, Lomis K, Johnson K, Miller B. Changing medical school IT to support medical education transformation. *Teach Learn Med* 2016;**28**(1):80–7.
20. Artino A, Brydges R, Gruppen LD. *Selfregulated learning in healthcare profession education: theoretical perspectives and research methods. Researching medical education*Oxford, UK: John Wiley & Sons; 2015. p. 155–66.
21. Hattie J. *Visible learning: a synthesis of over 800 meta-analyses relating to achievement.*Abingdon: Routledge; 2008.
22. Brooks C, Thompson C. *Chapter 5: Predictive modelling in teaching and learning*. 1st ed. Athabasca, Canada: Athabasca University Press; 2017 Accessed July 17, 2017.
23. Cunningham JA, Bitter G, Barber R, Douglas I. *Using traces of self-regulated learning in a self-paced mathematics MOOC to predict student success*. 2017.
24. Liu R, Koedinger KR. 1st ed. Lang C, Siemens G, Wise A, Gasevic D, editors. *Handbook of learning analytics*Athabasca, Alberta, Canada: Athabasca University Press; 2017.
25. Baker RS, Hershkovitz A, Rossi LM, Goldstein AB, Gowda SM. Predicting robust learning with the visual form of the moment-by-moment learning curve. *J Learn Sci* 2013;**22**(4):639–66.
26. Jiang Y, Baker RS, Paquette L, San Pedro M, Heffernan NT. Learning, moment-by-moment and over the long term. In: *International conference on artificial intelligence in education*; 2015.
27. IBM. *Watson for Oncology. IBM Watson for health*. Available from: https://www.ibm.com/watson/health/oncology-and-genomics/oncology/; 2017 [accessed August 12, 2017].
28. *Sense.ly. Sense.ly features*. Available from: http://sensely.com/features/; 2017 [accessed August 12, 2017].
29. Edwards RL, Davis SK, Hadwin AF, Milford TM. Using predictive analytics in a self-regulated learning university course to promote student success. In: *Proc. of the seventh international learning analytics & knowledge conference*. Vancouver, British Columbia, Canada: ACM; 2017.
30. Rodrigo MMT, Baker RS, Lagud MC, et al. Affect and usage choices in simulation problem solving environments. *Front Artif Intell Appl* 2007;**158**:145.
31. Baker RS, Corbett AT, Koedinger KR, Wagner AZ. Off-task behavior in the cognitive tutor classroom: when students game the system. In *Proc. of the SIGCHI conference on human factors in computing systems*; 2004.
32. Arnold KE, Pistilli MD. Course signals at Purdue: using learning analytics to increase student success. In: *Proc. of the second international conference on learning analytics and knowledge*. Vancouver, British Columbia, Canada: ACM; 2012.
33. Lomis KD, Russell RG, Davidson MA, et al. Competency milestones for medical students: design, implementation, and analysis at one medical school. *Med Teacher* 2017;**39**(5):494–504.
34. Veletsianos G, Reich J, Pasquini LA. The life between big data log events. *AERA Open* 2016;**2**(3) 2332858416657002.
35. boyd d, Crawford K. Critical questions for Big Data. *Inf Commun Soc* 2012;**15**(5): 662–79.
36. Selwyn N. Data entry: towards the critical study of digital data and education. *Learn Media Technol* 2015;**40**(1):64–82.
37. Frank JR, Danoff D. The CanMEDS initiative: implementing an outcomes-based framework of physician competencies. *Med Teacher* 2007;**29**(7):642–7.
38. Frank JR, Snell LS, Cate OT, et al. Competency-based medical education: theory to practice. *Med. Teacher* 2010;**32**(8):638–45.
39. Dillon J, Ambrose GA, Wanigasekara N, et al. Student affect during learning with a MOOC. In: *Proc. of the sixth international conference on learning analytics & knowledge*; 2016.

40. Azevedo R, Millar GC, Taub M, Mudrick NV, Bradbury AE, Price MJ. Using data vi-
    sualizations to foster emotion regulation during self-regulated learning with advanced
    learning technologies: a conceptual framework. In: *Proceedings of the seventh international
    learning analytics & knowledge conference*; 2017.
41. D'Mello S, Kappas A, Gratch J. The affective computing approach to affect measurement.
    Emotion Rev 2018;10(2):174–83.
42. San Pedro MOZ, Baker RS, Heffernan NT. An Integrated look at middle school engage-
    ment and learning in digital environments as precursors to college attendance. *Technol
    Knowledge Learn* 2017;**22**(3):243–70.
43. Pardo A, Siemens G. Ethical and privacy principles for learning analytics. *Br J Educ Tech-
    nol* 2014;**45**(3):438–50.
44. Campbell JP, Oblinger DG. Academic analytics. *EDUCAUSE Rev* 2007;**42**(4):40–57.
45. Greller W, Drachsler H. Translating learning into numbers: a generic framework for
    learning analytics. *J Edu Technol Soc* 2012;**15**(3):42.
46. Goggins SP, Galyen KD, Petakovic E, Laffey JM. Connecting performance to social
    structure and pedagogy as a pathway to scaling learning analytics in MOOCs: an explor-
    atory study. *J Comp Assisted Learn* 2016;**32**(3):244–66.

PART FOUR

Evaluating and sustaining Digital Innovations

CHAPTER TEN

What's in your medical education data warehouse? Interviews with 12 US medical schools

Boyd Knosp[a], Michael Campion[b], Johmarx Patton[c], Helen Macfarlane[d], Janet Corral[e]
[a]University of Iowa, Roy J. and Lucille A. Carver College of Medicine, Iowa, IA, United States
[b]University of Washington, Seattle, WA, United States
[c]Association of American Medical Colleges, Washington, DC, United States
[d]University of Colorado School of Medicine, Aurora, CO, United States
[e]University of Arizona Tucson College of Medicine, Tucson, AZ, United States

Chapter outline

A brief introduction to education data warehouses

This chapter shares recent work of an environmental scan of education data warehouses (EDWs) across 12 medical schools in the United States. The adoption of EDWs is driven in part by accreditation requirements to continuously monitor curricular program elements. Medical schools in the United States are devoting increasing resources to establish data infrastructures to facilitate access and analysis of this data. The chapter

Digital Innovations in Healthcare Education and Training.
http://dx.doi.org/10.1016/B978-0-12-813144-2.00010-6
Copyright © 2021 Elsevier Inc. All rights reserved.

begins by exploring the definition of EDWs in terms of their composition and rationale as a *warehouse* and not simply a *database*. Implementation issues and considerations, such as: data governance, interacting with multiple stakeholders, and data across systems will be addressed before the chapter moves to present the summary data and analysis of the 12 semi-structured interviews.

Defining education data warehouses (EDWs)

Much of how EDWs are defined is borrowed from the business world, where data warehouses (also called enterprise data warehouses) are a normal part of business intelligence operations. In business, data warehouses are central repositories of data from one or more sources, where data is then used for analysis and reporting to guide business decisions. EDWs, similarly, are defined by their function and content. The function of EDWs is to store education data and make it available for analysis to a range of stakeholders. When discussing the data storage in education, the term *warehouse* is purposely chosen over *database* to better communicate a place of storage *and* delivery of data. Some business data warehouses extend functionality to include visualization and/or analysis tools, although these operations typically happen downstream from the place of data storage.

The second element of EDWs is content. Content emcompasses two variables: *data domains* and *information technology systems*. Data domains are defined by the many types of education data that exist; for example: student demographics, student assessment, program/educator evaluation, learner interactions, external examination scores, etc. A selection of data domains for medical schools in the United States is depicted in Fig. 10.1. Information

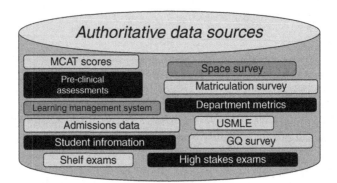

Figure 10.1 Data sources typically included in an Educational Data Warehouse.

technology systems are the computer-based systems that generate or maintain the live data; for example, the learning management system, online exam delivery software, the student information system, and lecture capture system. Information technology systems may also be external to institutions, including external standardized exam systems (e.g., MCAT, USMLE Step 1), national admissions application systems for undergraduate or graduate training, national graduate questionnaires, or third-party owned learning applications with curriculum or assessment content (e.g., online drill-and-practice question bank software, online videos of curricular content created commercially or through a consortium of institutions, etc.).

Scope of, and imperatives for, EDWs

The most basic of EDWs are those reporting on data from single systems (e.g., one learning management system, one online examination system), or manually merging data domains together in spreadsheets (e.g., merging institution-developed student exam results, from the MD program, with exam results from external standardized exams, such as the USMLE Step 1). The more complex EDWs consist of data from multiple domains and systems, which are fed into a dedicated database infrastructure (consisting of one or more databases, with live feeds or static, periodic updates). Importantly, the relatively limited budgets within the education missions of most health professions education programs often impact the choice of infrastructure (e.g., spreadsheet versus formal database architecture), the simplicity or complexity of the EDW, the speed of constructing an EDW, as well as the extent of data that can be stored and reported upon.

When discussing the scope of EDWs, it is important to delineate the EDW from related operational decisions or structures. One key operational piece, data governance, is used to: clarify and specify what data is in an EDW; when it is stored and checked for quality; how quality is confirmed; who has access to which data in the EDW; and defining specific conditions under which access is available. Data governance is best embarked upon as the first, or one of the early steps, in founding an EDW. A second key operational piece was alluded to previously, which is data analysis and visualization. Additional software, such as a statistical package or a visualization package, are used by stakeholders with access to the data in the EDW, but who do not necessarily analyze (i.e., edit or change the data) within the EDW itself. The EDW typically stores exact or cleaned copies of original data, and/or integrated data from multiple sources. Analysis and visualization usually happen outside of the EDW in order to maintain the integrity of the contents of the EDW. However, health professions education programs with larger budgets may have the means to incorporate automated analysis and reporting as part of their EDWs.

Lastly, the operational imperative for EDWs has varied over time. Medical schools, who were early adopters of EDWs, did so to provide infrastructure in order to complete experiments and analyses as part of continuous quality improvement within their evaluation of learners.[1] However, in the past few years, the reporting requirements for accreditation have become extensive, with the authors' professional experiences being upward of 2 years to collect, analyze, and report requested data. Simultaneously, the complexity of the data to be reported has increased. Providing reports for accreditation purposes has thus become a time and resource intensive process that could be more efficient if there was an established repository of accurate data that could be analyzed by those stakeholders needing to create reports. Thus, EDWs are quickly becoming an operational imperative for the contemporary medical school.

Yet, much is not understood about EDWs. Infrastructure, data domains, personnel, and operational uses of the data in an EDW are non-standardized. An environmental scan to understand implementations, data composition, and emerging best practices, is needed.

Environmental scan: methodology

Semi-structured interviews with 12 US institutions were conducted, between late 2017 and Spring 2018, to better understand the composition of EDWs and their implementation within medical schools. A range of medical schools participated, from those with established implementation to new adopters, and inclusive of private (n = 6) and public (n = 6) institutions. The interviews were conducted by at least two authors (sometimes up to five authors) with individual institutions, who were welcome to include as many participants from their school as they deemed necessary to answer the questions. Participants therefore often included members of the academic technology teams (n = 31), and also administrative leaders such as Assistant or Associate Deans (n = 6) or members of the Evaluation team in the Office of Medical Education (n = 4) Table 10.1.

Interviews lasted approximately 60 minutes and followed the semi-structured protocol (Appendix A, end of chapter). Authors took notes during the interview, which were posted to a common document hosted

Table 10.1

School (Private/Public)	Participant #	Participant role at institution
NYU (Private)	1	Associate Dean, Educational Informatics
NYU (Private)	2	Associate Director, Information Management, Chief, Education Data Section
University of Michigan (Public)	3	Director, Education Informatics
University of Michigan (Public)	4	Informatics Analyst/Program Manager
University of Colorado (Public)	5	Director, Educational Technology
University of Colorado (Public)	6	Director, Evaluation
University of Colorado (Public)	7	Manager, Education Data Systems
Emory University (Private)	8	Assistant Dean, Medical Education
Emory University (Private)	9	Deputy Chief Information Officer
Emory University (Private)	10	Assistant Dean, Medical Education Research
Emory University (Private)	11	Senior Director, IT
University of Iowa (Public)	12	Associate Dean, IT
University of Iowa (Public)	13	Lead Application Developer
University of Iowa (Public)	14	Director, Office of Curriculum Research in Medical Education
University of Iowa (Public)	15	Faculty, Office of Curriculum Research in Medical Education
University of Iowa (Public)	16	Assistant Dean, Education
University of Iowa (Public)	17	Project Manager
Mayo (Private)	18	Education Reporting and Analytics
Mayo (Private)	19	Program Manager
Mayo (Private)	20	Program Manager, Ed Reporting and Analytics
Mayo (Private)	21	Program Evaluator, Education
Texas Tech University (Public)	22	Associate Director, Academic Technology
Texas Tech University (Public)	23	Associate Dean, Medical Education
UC Irvine (Public)	24	Assistant Dean, Education Compliance and Quality
UC Irvine (Public)	25	Associate Dean, Clinical Science Education and Educational Technology
UC Irvine (Public)	26	Quality Data Analyst

(*Continued*)

Table 10.1 (*Cont.*)

School (Private/Public)	Participant #	Participant role at institution
UCSF (Public)	27	Associate Dean, Education
UCSF (Public)	28	Executive Director, Technology Enhanced Education
UCSF (Public)	29	Director, Data and Analytics
UCSF (Public)	30	Director, Technology Strategy and Development
UCSF (Public)	31	Leading Enterprise Data Analytics and Governance
University of Southern California (Private)	32	Director, Educational Technology
University of Southern California (Private)	33	IT Manager
University of Southern California (Private)	34	Computer Consultant
University of Southern California (Private)	35	Programmer
University of Southern California (Private)	36	Programmer
University of Washington (Public)	37	Director, Academic and Learning Technologies
University of Washington (Public)	38	Director, Project and Information Services
University of Washington (Public)	39	Director, Education Evaluation
University of Vermont (Public)	40	Chief Information Officer
University of Vermont (Public)	41	Senior Applications Developer

online (Google Docs, docs.google.com, Cupertino, CA, USA). The notes were reviewed by the multiple authors involved in the interview, who added clarification and edited notes from the interviews. The notes were reviewed by the participants from the interviewed institution before each interview summary was considered to be final. All authors analyzed the interview summaries using *a priori* themes. Themes and codes were recorded in a spreadsheet (Google Sheets, sheets.google.com, Cupertino, CA, USA). All authors reviewed the themes and codes for saturation, as well as to come to consensus on final themes and codes. The final themes and codes were organized and presented in a working group session at the Association of American Medical Colleges (AAMC) 2018 Information Technology in

Academic Medicine Conference sponsored by the AAMC Group on Information Resources (GIR).[2] The working group included many participants of the present study, as well as information technology and academic
technology professionals of similar rank and role at other institutions. Their
feedback thus constituted member checking and validation of the study
findings.

Results

A total of 41 members from 12 schools participated in the interviews. The average number of participants from each school was 3; there
was a range from 2–5 among all participating schools. The *a priori* themes
addressed: what data is in an EDW and where it is stored; how it is used;
resources (including infrastructure, people, and governance); and, future directions for EDWs within each medical school.

What data and where it is stored

There is a range of basic and advanced EDWs among participants. Seven
participants use a formal infrastructure dedicated to store original or accurate copies of data from one or multiple systems, for the purposes of storage,
analysis, and reporting. These schools were noted as "central aggregators,"
as their EDWs comprised a central server, central database, or central data
mart (i.e., a self-service portal for stakeholders to access EDW data). In contrast, five other participating schools noted their EDW data was distributed
across internal and external systems. For example, one school has exam data
housed on a cloud-based third party provider Software-As-A-Service examination system, where the exam data is pulled from and into the internal
electronic portfolio to track student progress. Other schools shared similar
examples of data hosted in third-party systems, including: clinical rotation
evaluation data, program evaluation data, and pre-clinical learner evaluation
data. Almost all schools participating in the study get a "data dump" (one-
time data pull) from external regional or federal systems (e.g., admissions),
which they store in their EDW. Specific platforms or software services mentioned by the participants in the interviews are noted in Table 10.2.

Three participants were working with what they considered to be simpler forms of EDWs. These simpler EDWs ranged from spreadsheets to
multiple spreadsheets, for storing data from one or more systems. Some
participant medical schools were converting a national survey reported in
Portable Document Format (PDF) form, into spreadsheets before the data

Table 10.2 Infrastructure for EDWs among participant medical schools.

SQL server, Tableau, Microsoft Excel, Microsoft reporting services
Using Tableau for student dashboard of exam scores

Leveraging University data warehouse, also pulling from Canvas (a learning management system), using SQL, Tableau and possibly SQL server reporting services (SSRS) in future, web portals, and Excel/SPSS/SAS (Note: SPSS and SAS are statistical software packages)

DB2 (a relational database management tool from IBM) for unified data platform; SAP for universe and cubes; Business Objects for reporting; Tableau for dashboards

Oracle DataWarehouse and Tableau

Tableau and central IT supports servers

Microsoft SQL, Microsoft reporting services, Tableau. Servers managed by central IT.

Microsoft, Sharepoint, Power BI, Peoplesoft

SQL, Power BI

Oracle DataWarehouse

SQL, ClickView and implementing Tableau

could be analyzed, or combined with internal school data for secondary analyses. The authors called this working with data "in place."

A secondary finding—the difficulty with data coming in "locked" formats such PDFs—raised a sentiment of frustration among participants. Participants shared the limitations of "locked in" data sources, that is, these sources were formatted or already analyzed, rather than providing the original data from which stakeholders would conduct multiple analyses. The "locked in" nature also prevented teams from being efficient since many hours and/or team members were involved in manually copy/pasting values from PDF reports into other document formats (e.g., spreadsheets) or software (e.g., statistical software). Participants vocalized a growing need for standardized or interoperable data standards among EDWs, both internal and external to medical schools, to facilitate efficient data collection, aggregation, analysis, and reporting.

How data is used?

There are three main ways in which the medical schools interviewed in this study use data within their EDWs: tracking student progress throughout program; student success, including research and early detection; and, for accreditation purposes and for continuous quality improvement of the curricular program.

Tracking student progress throughout the program was the top reason for data usage within an EDW. Five schools used student assessment data from their EDW to report learner progress to visualization dashboards available to learners, faculty coaches, and/or program administrators. One school has a high proportion of undergraduate learners who stay on for postgraduate training, and thus their particular EDW works for longitudinal tracking. One school uses EDW data for monitoring students' clinical rotations in the third year of their undergraduate training. One school allows educational researchers for accessing EDW data to complete inquiries and analyses related to medical education research projects.

Student success is a related function that encompasses various efforts among the multiple participant schools in the study. One school uses their EDW data to calculate the extent any particular learner is at risk of requiring remediation. Another school uses EDW data to evaluate how well the admissions process is working. Three additional schools share EDW data directly with students through a dashboard, allowing learners to view their education profile including a data comparison of the individual with their cohort of classmates. One school provides financial coaching using EDW data, and one school uses EDW data within student advising.

Nine of the twelve schools interviewed explicitly stated their EDWs were for accreditation and/or continuous quality improvement purposes. Accreditation of medical schools in the United States and Canada, is done through the Liaison Committee for Medical Education (http://lcme. org/)*. LCME Accreditation requires compliance with 12 standards across three levels: the institution, the program, and the learners (http://lcme.org/ publications/). In particular, standards 6, 7, and 8 relate to the curricular structure and quality, and require a significant number of reports drawing from multiple data sources (e.g., assessments, admissions, faculty ratings of learners, etc.). The first accreditation standard also requires continuous quality improvement (CQI) of the MD curricular program as a whole. CQI involves monitoring objective outcome measures to ensure short and long-term program goals are met. CQI also involves ongoing monitoring to ensure accreditation standards are met over time. Both CQI activities benefit from EDWs as reliable storehouses of accurate data drawing from the multiple sources needed for accreditation reporting, and for creating reports in time and resource efficient ways.

EDW resources: people and governance

People

Only one participant medical school has a dedicated and formal EDW team of three people. For 10 of the participating schools, it was more typical to see personnel from other teams spending part of their time (FTE or Full Time Equivalent) on the EDW work and initiatives. At least two schools responding have dedicated people to conduct analyses. One school noted that their team was informal. This finding is important as it relates to both the emerging nature of EDWs at medical schools, and the need for committed funding to continue efforts in earnest. For a breakdown of roles and percent effort by school, please see Table 10.3.

Table 10.3 Personnel involved in EDWs.

Three FTE on EDW team

One dedicated FTE (data analysis), Two partial FTE (data mgt/dev); tech support in Deans' office (servers, etc.)

Medical education/student affairs, IT all provide partial FTEs for a total of 1.2 FTE effort on our EDW

Four IT staff, multiple people within the Education office

Education: Three primary people partial FTE; IT: Four dedicated FTE plus some partial FTE (PM, Data modeler, DBA, ETL programmer, report developer)

2.75 FTE split across 4–5 people in: education and clinical Deans' offices and Central IT services

About twoFTE spread over 5 people. We have central IT support for Tableau and server infrastructure

Deans and Directors in SOM are involved. There are a team of 4–5 that provide 1–1.5 FTE for this effort.

Informal group, about seven FTE

Less than five dedicated FTE, "understaffed"

Five people, but two FTE

Three groups, no dedicated people but portions of up to eight FTE

Close to two FTE

Six FTE across teams

Note: "FTE" stands for full time equivalent. A whole number indicates 100% time on the EDW (e.g., 3 FTE = 3 people working 100% of their time on the EDW).

Governance

Data governance efforts are at various levels of maturity across the 12 schools interviewed. Table 10.4 outlines their efforts and composition, where available. Notably some schools are proceeding with live EDWs without formal governance in place, often under pressure to complete reporting for accreditation. There is no single common composition of data governance committees; in three schools, there are members from senior administration in either the education mission or the clinical enterprise. In one case, the Director of Information Technologies is the sole decision maker.

Future directions for EDWs

There are two main directions that participant medical schools noted as important in the coming years. The first is their immediate "next steps" or

Table 10.4 Governance.

Learned hard way with *ad hoc*, then developed governance committee modeled on clinical and research with written policy document that is reviewed annually

Governance committee chaired by Director of Education Technology. Governs decisions on security, what to include, data use, conflicts on data needs; assign work groups to resolve issues

None in place now—Director of IT and deans make decisions

"Data governance lite," need to define what data means (definitions), access, and using data to drive decisions

Leadership and administrators from five schools provide leadership; superuser group with representatives from all schools meets weekly

Governance—work in progress but they have Vice Dean of Education and Hospital providing leadership on data in their areas. An education and clinical dean is on EDW team

No formal governance structure, IT shop does most of the oversight

On hold until accreditation work completed

Formal governance structure with stakeholder representatives; have formal policy which is reviewed annually

Currently informal governance through leadership, looking at formal governance models

Building governance now

Data governance committee with IT and education people on it, 6 years old, funding also influences where the work is done "looking ahead" goals toward making a more robust EDW locally. Given the range of maturity of EDWs in the present study, there is a range of "next steps." Eight schools focus their next steps on better reporting to learners, or more robust users of learner data. This underscores the mission of EDWs as focused on learner success. Two schools mention machine learning as an important future development, while two other schools noted the more immediate importance of overcoming barriers with vendors, such as access to data.

The second main direction of EDWs is the sharing of data. For half of the participants in the interviews, sharing data was intra-institutional (i.e., across colleges, schools or administrative units within their own schools). For many participants, inter-institutional data sharing is seen as a key future goal. There were many applications for inter-institutional sharing, including answering educational and health care workforce research questions, educational benchmarking, and administrative benchmarking. Concerns persisted about linking data accurately, as well as maintaining an accurate central repository within institutions.

One school, an early adopter of EDW, responded that they sought transformative change through the use of their EDW. This comment represents the promise that draws schools to continue to labor in an area that is routinely under-resourced.

Discussion
Variability among EDWs is high

Notably, there is not one common EDW infrastructure across participant medical schools, nor are the data domains common to all EDWs in the study. Thus, the variability among EDW structure and data domains is highly variable; in fact, participants at one medical school do not currently have an EDW.

EDWs are not uniformly sustainable—yet

Multiple schools in this study reported that they could not afford formal information technology infrastructure. Of those that did, several funded the original start-up with grant funding and/or one time infrastructure investments. Simultaneously, multiple schools also reported the time and resource intensive nature of having to compile data manually. Taken together, these findings have ramifications for the sustainability of EDWs:

1. There is a need for validated maturity models as well as documented business cases explaining the competitive advantage provided by EDWs. Informatics leaders need such documentation to advocate for long-term planning and budgetary requests to support robust EDWs.

2. The medical education community—including external stakeholders— need to commit to common data standards and/or reliable methods of connecting to external reporting authorities (e.g., through Application Programming Interfaces (APIs)) in order to maintain the efficiency behind EDWs. Both bringing external data into an EDW, and reporting out of an EDW to external authorities, are time consuming processes for individual medical schools.

The higher education community, while small and typically underfunded, has several precedents that may help bring solutions to both aforementioned needs. First, EDUCAUSE, a not-for-profit organization based in the United States (https://www.educause.edu), has produced multiple reports and one framework for maturity models (e.g., Dalhstrom[3]). These have previously been adapted for academic technologies in medical education. Second, common data standards have been created and supported by several groups. MedBiquitous, a not-for-profit based in the United States (https://www.medbiq.org/), has helped create and maintain data standards (e.g., ANSI Virtual Patient Standard). Similarly, the advent of non-profit collaboratives like IMS Global create and provide interoperability standards for moving learning objects between informatics platforms, and may be a source of data interoperability standards in the future.

Limitations of the study

While the study did purposefully sample public and private medical schools, it was exclusive to schools in the United States. Including international schools, may reveal a wider range of practices and lessons learned.

The interview methodology was semi-structured. This permitted the interviewers and the interviewees to explore local contextual needs and/ or issues. This approach, while reasonable for an environmental scan, needs additional rigor in order to create best practices with EDWs.

Differences in privacy laws, student data protection laws, public records laws, and reporting to government Education or Health agencies were not explored in this study, as the focus was on understanding the composition of EDWs. As such, the findings of the present study have limited application in other countries or districts where local, regional, or federal regulations and laws impact the composition and function of EDWs.

Future considerations for EDWs and learning analytics

The landscape of what is in EDWs is rich. Future studies would be wise to consider validating the findings of the present study in a larger, international community. Such a study might also yield more extensive lessons learned, as well as a better appreciation for the range of contexts, legalities, and regulations that impact EDWs across medical schools internationally.

One future possibility for EDWs is the inter-school and inter-institutional reporting of data in a more efficient manner. The present study found that interoperable data standards are something yet to be achieved, and greatly needed in order to achieve this goal. The community at large will need significant investment in existing standards development (e.g., Med-Biquitous, https://www.medbiq.org/; EdMatrix, https://www.edmatrix.org) in order to reasonably consider a future where efficient reporting to governing institutional bodies directly from an EDW would be possible.

Conclusion

EDWs are foundational infrastructure investments for medical schools in the modern era. The range of EDWs is strongly related to the internal resources of individual medical schools, and often involves leveraging internally available software and hardware in order to exist and to continue to exist. Yet, the modern business world outside of medical schools continues to utilize data warehouses as a normal part of operations, with artificially intelligent and/or automated analysis being a key part of each company's competitive advantage. Medical schools are wise to make the investment into EDWs, as well as to consider collective efforts to create and adopt interoperable data standards to make internal and external reporting—currently a time and resource consuming process—more efficient.

Appendix A   Semi-structured interview

The following questions constitute the semi-structured interview in the present study. Authors who conducted the interviews were permitted to ask follow-up questions based on the participants' responses. If follow-up questions were asked, they were usually for clarification or to better understand the local context.

1. What data and what are you doing with it?
2. Who are your target audience and uses?

3. What resources (people and tech) support Med Ed Data?
4. What are your biggest challenges?
5. What is your vision for the future for your EDW?
6. What is your vision for the future for EDWs broadly in academic medicine?

References

1. Stuart G, Triola M. Enhancing health professions education through technology: building a continuously learning health system. *Proc. of an AMA conference recommendations.* Available from: http://macyfoundation.org/docs/macy_pubs/Macy_Foundation_Monograph_Oct2015_WebPDF.pdf; 2015.
2. Campion M, Corral J, Kendrick E, Macfarlane H, Knosp B, Patton J. What's in your Med Ed data warehouse? Leveraging data to support the educational mission. *Presentation at the American association of medical colleges group on information resources conference.* June 5–8; 2018: Austin, TX.
3. Dahlstrom E. Moving the red queen forward: maturing analytics capabilities in higher education. In: *EDUCAUSE Rev*; 2016.
4. Calzada J, Knosp B, Mantia J. Maturity and deployment indexes in medical education technology. In: *2015 AAMC information technology in academic medicine conference,* San Diego, CA; 2015.

CHAPTER ELEVEN

Teaching and integrating eHealth technologies in undergraduate and postgraduate curricula and healthcare professionals' education and training

Eirini C. Schiza[a], Maria Foka[b], Nicolas Stylianides[b], Theodoros Kyprianou[b], Christos N. Schizas[c]
[a]Department of Intelligent Systems Group Johann Bernoulli Institute for Mathematics and Computer Science University of Groningen, Groningen, The Netherlands
[b]Department of Intensive Care Medicine, Nicosia General Hospital, Nicosia, Cyprus
[c]Department of Computer Science, University of Cyprus, Nicosia, Cyprus

Chapter outline

Digital Innovations in Healthcare Education and Training.
http://dx.doi.org/10.1016/B978-0-12-813144-2.00011-8

Introduction

The use of "eWords" such as eBusiness, eCommerce, and many more, is an attempt to convey the principles, enthusiasm, and high expectations that the electronic era and the ever-expanding use of Internet tools has created. This new reality shaped new opportunities, addresses, and creates numerous challenges to the traditional healthcare information technology industry and coined the use of a new term eHealth. Several definitions have been used in academic literature for eHealth, such as medical informatics or information and communications technologies for health (ICT for health).

From the economic perspective, eHealth comes to increase efficiency in an industry which exhibits one of the largest inflations in employment, that is, healthcare, thereby decreases the costs by eliminating duplication of tests, unnecessary examinations, and admission through enhanced communication possibilities between healthcare institutions and the patient involved. Thus, we can improve quality, by allowing sharing of medical data between different healthcare providers and directing the patient to the best, more suitable, and available medical facility always and at all places. Effectiveness and efficiency are guided by evidence-based scenarios. Evidently, from the healthcare perspective, eHealth revisited the relationship between the patient and the doctor by enabling change of the classical statement "the doctor will see you now" by "the patient will see you now,"[1] forming a new, patient-centered philosophy.[2] Diagnostic and therapeutic decisions nowadays take place in a shared data environment thus their accuracy and efficacy are vividly improved.

Technology is meant to help the patients, the citizens at large, and healthcare providers. Students studying medicine, for instance, need to be aware of the eHealth technological developments as their future working environment is rapidly and continuously changing now and, in the years, to come. Medical faculty in turn, needs to prepare students for a successful entry into the practice. Even though the first-year student these days are much more computer literate when compared to the students few years ago, we should not entertain our fears of the dangers of staying behind from making full utilization of the opportunities that ICT can offer to healthcare and healthcare education. By the time a first-year medical student completes one's studies and specialized training and end-up practicing medicine at least 8 years will lap. As the evolution to ICT is continuous, 8 or even 3 years bring major changes that influence the profession. Therefore, the medical student should be taught not only today's trends in using ICT means but to learn to invent the future and vision the future with an open mind. Nowadays, academics teaching eHealth to the medical student should inspire them to invent the future in medical practice and similarly provide their students with competences to use digital innovation to keep up-to-date with the latest advancements in their area of practice. In addition to this, the eHealth academics should teach eHealth in an entirely different way when it comes to postgraduate level, and entirely different when it comes to the medical practitioners in hospital and clinics at present. The courses for the last category should be adapted in a case by case manner in order to accommodate the needs of medical practitioners who entered practicing a year ago at one end and those who entered medical practice decades ago on the other end. The answer to this reality is training at the job using digital innovations in a continuous way to enhance one's skills but most importantly for preparing oneself for the present and the future.

Stated in a philosophical way an eHealth academic has the imperative responsibility to inspire students and medical practitioners to appreciate the motto "I am aged as a learner – Διδάσκω αεί διδασκόμενος," the paternity of which belongs to Solon (640–560 BC), the well-known Athenian lawmaker and philosopher. And what can bring state-of-the-art knowledge on the go better than digital innovations for education and training?

Learning methodology: Evidently, teaching and learning methods at undergraduate, postgraduate, and professional levels can be very different. To promote eLearning approach, it is essential for institutions to introduce and support the use of eLearning technologies at all levels, from learning resources, to teaching and assessment. This is especially important for postgraduate students that use a more self-directed style of learning. Professional education and continuous education pose nowadays important challenges (i.e., time, style, motivation, and resources limitations) and eLearning may offer solutions to those challenges. It may do so by integrating learning processes into the work environment, by changing teaching style to microlearning and competency-based learning, thus stimulating and incentivizing professionals.[3,4]

First, we describe a model of preparing medical students for the afore-mentioned challenges, in a two-semester introductory course on eHealth, placed in their first year of their undergraduate studies. The curriculum aims to introduce students into eHealth technologies and competencies and give them the theoretical background and the exposure necessary to understand these applications and their clinical, ethical-legal, and cost/effectiveness characteristics. Then we describe an innovative curriculum for a newly established, part time, distance learning MSc in applied Health Informatics & Telemedicine tailored to mixed cohorts of healthcare/IT graduates. It offers eight modules as follows, (1) National Health Information Systems & Hospital Information Systems; (2) Standards & Technology assessment; (3) Patient Medical Records and Electronic Prescription Systems; (4) Medical Image processing and analysis; (5) Telemedicine applications; (6) Patient data management & decision support; (7) Biomedical databases & Biomedical research methodology; and (8) Tendering and Hospital Information project management—Design to deployment. The technical details being a total of 100 ECTS spread to: 10% live/interactive sessions, 70% self-directed learning and assignments, virtual labs (Image processing, Data analysis, Medical Record, Tendering & eHealth Project Management, and Health Informatics Standards). The last course to be presented was developed by the Nicosia General Hospital-Intensive Care Unit and the Open University of Cyprus, in collaboration with the University General Hospital-Intensive Care Department of Heraklion in Crete. This collaboration was co-funded by the EU-INTERREG Cross-Border Cooperation Program "Greece–Cyprus 2007–2013" Tele-Prometheus. This program is aimed to revolutionize continuous education/learning process in ICU healthcare professionals and patients' families. Emphasis is put to the use of eHealth and digital innovation tools to facilitate learning processes and optimize learning outcomes and describes success indicators in learning and practicing eHealth. Triangulation of perceived educational needs, peer performance review, and clinical audit results is described as an innovative approach to personalized learning, especially at the continuous professional education (CPE) level.

It should also be mentioned that in this book chapter we introduce a continuum of eHealth education modules/services taught, at the undergraduate level, at a post-graduate level distance learning, an Applied Health Informatics & Telemedicine MSc course, and at the professional/patients' education level, embracing the general principles shown in Fig. 11.1 and 11.2.

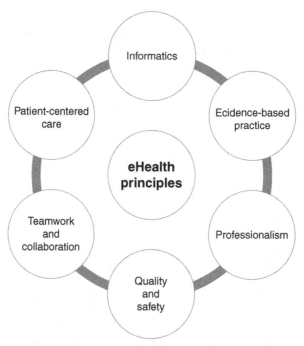

Figure 11.1 *eHealth Principles.*

Figure 11.2 *TelePrometheus development methodology.*

eHealth education curriculum

Undergraduate level

Module: electronic health and medical informatics[5]

AFMC (The Association of Faculties of Medicine of Canada)[6] has carried out a project with the goal to improve clinical practice and patient care by supporting clinicians adopting and using electronic health record systems. Based on their findings, topics to be considered for inclusion in an eHealth training curriculum could be:

- Patient physician communication and professionalism, privacy, confidentiality, ethics, and rules of accessing patient records.
- Information literacy, information management, searching for information, web databases, critical appraisal, and evidence-based medicine.
- Specific instruction on local EHR or similar systems.
- Different types of technologies and records used.
- eHealth in other specific applications (i.e., Telemedicine systems).
- Using mobile devices.
- Medical imaging and imaging management systems, such as PACS.
- Mobile health.
- Social media for use in discussion and information sharing.

One of their major concerns when this was proposed is that none of Canadas' Medical Faculties teaches eHealth as a separate autonomous course, a fact that may pose difficulties to the students adopting and using effectively eHealth technology later in their professional training and most importantly using it constructively in practicing medicine during their career.[7] Our preposition which was applied few years ago was to introduce two eHealth courses during the first year of studies, one for covering the theoretical aspects with a dynamic syllabus based on the earlier-mentioned topics and extended, followed by another course in the form of seminars during the second semester of studies. The latter to be administered by selected medical practitioners who are relatively ahead of their time and as pioneers have employed eHealth principles in their everyday routine practice.

Learning objectives and curriculum

The course CS041 is taught during the first-year of a 6-years medical curriculum and before the students are exposed to the clinical environment. This aims to ensure that they acquire knowledge and skills for understanding better and appreciating the new eHealth technologies. The content of the course is dealing with the legislative, social, and ethical issues of eHealth, helps students to understand and use patient-centered approach in their medical practice. It also introduces students to medical practice/mechanisms of health and disease modeling, data processing, and new knowledge generation helping them to classify, standardize, and use biomedical health data for preventing diseases, following the principle "invest in health to avoid illness." Additionally, the principles of data acquisition, management, standardization, and presentation of information are taught. The objectives and the content of the two courses is analyzed later.

Digital innovation enhancing teaching and learning

The course has a dynamic and interactive teaching environment, which is provided by the Moodle platform. This innovative tool can facilitate mid-term and final examinations with open notes, grammar correction, and antiplagiarism tools. All exams are administered with open notes and internet access. By the end of the semester, students created working groups and worked on a project where they uploaded their final work and presentations. Digital innovation does not only improve teaching and learning methods but creates also an interactive and participating environment which encourages teamwork and modern ways of teaching and learning for teachers and students respectively.

Semester one (Course CS041: eHealth and Medical Informatics)

During the first Semester of their studies, a lecture series is followed in a dynamic Moodle environment with student participation using their laptops for accessing real-time teaching material as directed by their professor. An electronic class is created, and the students have the lecture notes and other relevant material, do and submit their homework online, generate discussion groups, and take their examinations online.[5] Selected topics are outlined later even though the topics may change from year to year due to the dynamic change in technology and its penetration in the medical profession. Most of the notes are likely to become outdated the next day and thus a dynamic teaching environment, such as Moodle helps the instructor and the students to keep up with the changes and the new technological achievements penetrating the medical practice in many aspects. In addition to technology, there are relevant legal issues and European directives which change and must be followed. The student must become aware that the medical professional of the future must be fully computer literate and be continuously updated. This will safeguard that technology will not overcome the medical professional; technology should be the servant of the user. The myth that technology will substitute the human in the medical practice should remain a myth. It's a truth which will never change that: technology is a good servant but a terrible master.

- eHealth: philosophy—science—technology

The first lecture is about the history of eHealth. Introduction to the basic concepts of information technology in health and the basic concepts of computer systems with reference to health applications and description of medical information exchange scenarios.

• Forms of eHealth

These two presentations focus of what is the meaning of eHealth and how European Union and others defines it. Some examples of the use of eHealth are presented like telemedicine, consumer health informatics, mobile health etc. Also, an overview of a National Health System (NHS) and the necessary support information systems both at national and European levels are presented. Analysis of the European Directive for Patient Summary for enabling the cross-border healthcare is given.[8]

• Medical technology and Information technology

Explanation of what medical technology represents through the years and how this evolved. Legacy systems such as Laboratory Information System (LIS), PACS, etc., are presented. Furthermore, Radiology Information System (RIS), that is, the analysis and description of the basic principles of systems that support imaging systems. Principles of interfacing and communicating with the electronic health record (EHR) support subsystem and the record management system (RMS). Finally, various management system activities such as the analysis and description of the basic principles of the information system for hospital's wider administrative and financial services (accounting, human resources management, materials and consumables management, and warehouse management) are presented.

• Electronic health record: International standards and applications of the EHR

Analysis and description of the basic elements included in the EHR. These elements are the patient summary and the extended patient summary which are both well-structured and compulsory and an optional part which is used mainly by the owner as notepad for exchanging notes with his doctor. Emphasis will be given in explaining how interoperability among health professionals is achieved by a well-structured scheme. A fully covered healthcare system both at national and European levels can only be achieved if a standard and interoperable EHR becomes available. In this sense, EHR can be characterized as the cornerstone for building a national healthcare system[2].

• Mid-term examination

Online examination is administered with open notes facilitated by the Moodle platform.

Table 11.1 Summarizes the digital innovations that used and can further advance the Semester one (Course CS041: eHealth and Medical Informatics).

Topics	Digital innovation used for teaching and training	Digital innovation that can further advance teaching and training
eHealth: Philosophy—Science—Technology	Dynamic teaching environment such as Moodle	Gamified applications to ensure understanding of the key concepts.
Forms of eHealth	Interactive presentations	Students implement scenarios based on a simulated environment for telemedicine, consumer health informatics, mobile health applications, Laboratory Information System, electronic health record, support subsystem, and the record management system in order to enhance experiential learning.
Medical Technology and Information Technology		
Electronic Health Record: International Standards and Applications of the her		
Mid-term examination	Online examination is administered with open notes facilitated by the Moodle platform	
Computational Intelligence and Medical Diagnostic Systems		Use of reusable learning objects to reach the learning outcomes of the topic through these bite-sized multimedia interactive online resources.
Legislative and Social Framework for eHealth—The European Experience		Use of game-based scenario to make links between the legislation for the management of medical information and practical cases.
eHealth Vision and National Healthcare Systems		Use of blog/wiki to collaborative adapt a legislation to a local context to identify the challenges of such procedure.
Group Presentations by Students	Visual aids uploaded by the students	Use of social media, Moodle forum to enhance collaborative learning.
Final Examination	Is administered with open notes facilitated by the Moodle platform	

- Computational intelligence and medical diagnostic systems

Computational intelligence for building diagnostic systems helps to capitalize and build on the vast amounts of medical data that exists in the EHR databanks. Through appropriate anonymization and data protection methods and technologies, biomedical data cohorts become available for researchers to develop intelligent tools for early diagnosis, prognosis, and thus prevention. Furthermore, telemedicine applications, image and signal processing applications are explored.

- Legislative and social framework for eHealth—The European Experience

It is explained why each country must make their own legislative framework based on the EU directives. Emphasis is given to the national and EU legislative system and problems of cross-border care. Reference is made to the key issues (medical confidentiality, medical information circulation, patient-centered approach versus doctor centered approach, patient rights, and health professionals). This section also covers interoperability issues and how proper legislation can help achieve it.

The European Interoperability Framework is presented through the *Integrating the Healthcare Enterprise (IHE)* protocols.[9] The open consultation framework for activating health professionals, suppliers, software developers, and healthcare solutions providers is presented from its legal perspective. Presentation of the use of computerized systems for the improvement of the medical services provided will be given and the legal issues associated are outlined.[10]

- eHealth vision and national healthcare systems

National roadmap and strategic pillars for evolving eHealth based on EU directives, adapted to the local environment is presented. They include public hospitals' reform and autonomy, quality assurance of services and efficiency, active research and development participation, e-Prescription at national level, and homogenization of public and private health institutions and services.[11]

- Group presentations by students

The students are taken in small groups to planned visits in a hospital and observe the data collection activity and follow the information flow system in a real medical environment. The students are given the opportunity to talk to the medical professional on the job and try to identify the decision-making activity and storage and retrieval of data. As a group exercise, students are asked to prepare a report with visual aids, upload on Moodle, share it with the rest of the class, and make a power point presentation in class to the other students. A discussion is followed and the whole session is video recorded for future reference.

- Final examination

Online examination is administered with open notes facilitated by the Moodle platform.

Semester two (Course CS042: eHealth and Medical Informatics Seminar Series)

During the second Semester, the course is turned into a Seminar Series where selected experienced doctors and other health professionals who are using eHealth in their everyday practice are invited to give a seminar based on their experiences. A selection of seminars on eHealth and Medical Informatics are given by selected healthcare providers. These seminars are aimed to reflect the everyday experience of the presenters in the use of eHealth technologies for making their practice easier and more cost-effective for their patients and institutions.[5] In this way, the students are given the opportunity to hear the truth and be advised by one of them to become (medical professional). Furthermore, the students are asked to select one of the topics presented and carry out a study by searching additional material from the internet and from sources given by the corresponding topic and presenter. Every student must take a step further and analyze this topic deeper, envision potential benefits, and explain how they see it applied in 5 years from present. They are also asked to argue on how an eHealth environment will change the life of medical professionals and citizens, and how societies will be reformed in the chosen area of medicine. The students at the end of the course will have to present their studies in front of the class and communicate their report to the professional who gave the respective seminar. The whole procedure is facilitated by the Moodle environment and as it progresses the students begin to appreciate more the interactive and participating teaching environment created. The course is being taught for the last 6 years and one can observe its dynamic change in the topic selection, content, and the ease by which one can find medical professional who are willing to present their professional specialization and show how this is continuously affected by the use of eHealth in the medical profession. Most importantly, the students come up in many occasions with ideas that can take one step further for the improvements seen and presented. This is turned into a very effective didactic tool because it triggers the imagination of the students and their curiosity for the medical profession in years to come.

The interaction of the presenters with the students however becomes beneficial to the former because occasionally during the discussions followed in the class or the discussions taking place via the Moodle environment, some very inspiring issues arise by the students which turn out to be beneficial to the presenters for further developments. Some of the seminar series titles given are:

- Electronic Health Record: From theory to practice—Examples from Internal Medicine and Infectious diseases.
- Electronic Health Record of the new-born and primary care: Pediatrics of the next generation.
- Informatics Applications and eHealth in everyday life, medical practice for optimizing therapy and safety of the patient.
- Electronic Health Records in the era of Value driven care.
- Electronic Health Records of total arthroplasty in Orthopedic Surgery: Scientific, clinical, economic, and political importance.
- Digital Technology as another invention in the service of medicine.
- Optimizing health care outcomes via Electronic Health Record.
- eHealth in Cardiology.
- eHealth and Precision Medicine.
- Bioinformatics: An essential factor in enhancing medicine in the direction of molecular precision and personalization.
- eHealth and Non-Invasive Prenatal Testing (NIPT).
- Biosignals in novel medical practice.
- Transport and management of patients with spinal cord injury within 12 hours is achievable by means of electronic medicine.
- The importance of eHealth in Ergonomics and Labor medicine.
- Generation—monitoring—and management of biosignals of the severely ill person in an ICU environment.
- The impact of modern technology in the prevention of heart attacks and stroke.

Teaching methodology

Lessons are taught using the Moodle platform, and include: Lectures/presentations (2 hours per week), tutorial (1 hour per week), and discussions. Visits to hospital clinics are made for observing the data flow, collection, and storing. Students are given the opportunity to explore how data is used by the medical professional for monitoring the patient's statues of the past, present, and future, which makes the appreciate the terms, prognosis, diagnosis, preventive medicine, precision medicine, etc. During lectures, students are required to have their personal computer for online searching and viewing as directed by the instructor. As part of the course during the second semester, at least eight medical professionals who are using eHealth in practice are invited according to the subject and their availability, from Cyprus and abroad for giving presentations to the students. It is also planned as mentioned earlier and when circumstances allow, accompanied by the presenter on-site visits of students to medical units for real time experience.

Digital innovations can enhance teaching and learning

The course CS042 has become beneficial to the student because the interaction and knowledge exchange was easier between students and teacher. Students had the option to revue at their pace the recorded presentations and the active discussions that was taking place during each presentation. The Moodle platform as mentioned earlier offers online final examination with open notes.

Assessment methodology

Student evaluation performance is based on individual work (exercises) (10%), teamwork (group exercises) (15%), midterm online examination (25%), and final online examination (50%).

Postgraduate level
Applied health informatics[12]

The curriculum exposes students to a comprehensive, high specification, and modern analysis of the various applications of computer science and web, PC, mobile applications in healthcare services. Skills development: (1) understanding of the most common problems and challenges in delivering effective and cost-efficient healthcare; (2) parameterization and analysis of relevant sustainable solutions; and (3) implementation and integration of the proposed solutions in existing clinical environments and monitoring of results and patients' outcome. Investment in knowledge, understanding, and skills development is carried out through virtual workshops and labs (https://www.med.auth.gr/sites/default/files/news-attachments/efarmos-meni_pliroforiki_igeias_tileiatriki.pdf).

Learning objectives and curriculum

National Health Information Systems & Hospital Information System.

Description: This course introduces the student to the sources of public health requirements such as knowledge of population size and characteristics, the cause of mortality and morbidity, and the state of health practice in a community. The course will also equip students working in a health information system environment with a fundamental knowledge of concepts and components of hospital information systems. The course will involve lectures, lab problems, and an end of term examination.

Learning objectives

1. Identify the state-of-the-art Health Informatics (HI) Applications and the need of introducing innovative technologies in Healthcare environments.
2. Explain the exploitation and application of HI.
3. Describe the fundamental concepts of contemporary technologies.
4. .29+ and their applications in the Health domain.
5. Summarize the importance of HI and standards in Healthcare applications.
6. Locate organization, technology, and institution problems and restrictions in HI applications.
7. Evaluate the different European and International Health Systems design approaches.

Standards and technology assessment

Description: This course analyses the present system of identifying and testing medical technologies and of synthesizing and disseminating assessment information. The course focuses on the flow of information that is central to an efficient assessment system. Methods for testing technologies and for synthesizing information are explored, and a compendium of data and bibliographic sources are included. The course also describes the innovation process for medical technologies, the effects that federal policies have on that process, and the needs those policies generate for technology assessment information.

Learning objectives

1. Identify the application of HI standards.
2. Explain the implementation of HI services using standards and the importance of communication interfaces among digital systems.
3. Describe the fundamental concepts of Coding and Taxonomy.
4. Summarize the needs of HI applications and the most common standards applied in HI systems.
5. Locate organization, technology, and institution problems and restrictions in HI applications.

6. Evaluate HI standards and the need of enforcing the usage of standards in HI systems.
7. Analyze the benefits of standard compliant HI systems.

Patient medical records and electronic prescription systems

Description: This course introduces the student to the important role of electronic medical records in today's complicated healthcare environment in planning, evaluating, and coordinating patient care in both the inpatient and the outpatient settings. The course completes the medical record model with an introduction to ePrescribing systems and issues related to pharmacy automation, medication compliance, pharmacy databases, CPOE, and adverse drug events (ADE).

Learning objectives

1. Identify the most important Medical Record applications
2. Explain the exploitation of standard compliant MR applications
3. Describe the fundamental concepts of MR applications.
4. Summarize the benefits of utilizing MR applications in Healthcare environments.
5. Locate organization, technology, and institution problems and restrictions of the MR applications.
6. Evaluate the most important MR implementations (open source vs. commercial vs. proprietary software).

Medical image processing and analysis

Description: This course gives an overview of medical image foundation enhancement, analysis, visualization, and communication as well as their applications in medical imaging. Basic approaches to display 1-D, 2-D, and 3-D biomedical data are introduced. As a focus, image enhancement techniques, segmentation, texture analysis, and their application in diagnostic imaging will be discussed. To complete this overview, storage, retrieval, and communication of medical images are also introduced. In addition to this theoretical background, an overview of useful software tools is given.

Learning objectives

1. Identify state-of-the-art tomography depiction technologies and the importance of tomography depiction in diagnosis and treatment.
2. Explain the technologies applied in biological tissue depiction.

3. Describe the methodologies applied in medical image reconstruction and the techniques in medical image enhancement.
4. Summarize the medical image analysis and processing procedure phases and the most important techniques used in every processing phase.
5. Describe the state-of-the-art storage, retrieval, and communication techniques in Medical Image Management systems.
6. Acknowledge state-of-the-art medical image automatic (or semi-automatic) processing systems and the importance of these in diagnosis.

Tendering and HI project management—Design to deployment
Description: This course examines traditional project management approaches to understand how they can best be applied to health informatics. A range of advanced technologies are introduced. Good practice in IT project management is examined and relate to the health context. A real case tendering example is used in order to better understand the tendering procedure.

Learning objectives
1. Describe the basic methods in management and administration of Health Informatics projects.
2. Evaluate the institution's required resources to develop a HI project.
3. Design, analyze, and resolve scheduling networks and identify project's critical tasks.
4. Propose effective human resource management methodologies to amplify effectiveness.
5. Risk assessment and Risk management.
6. Project Management.
7. Prepare and manage procurements.

Telemedicine applications
Description: This course provides a review of the history of telemedicine and telemedicine applications. Comprehensive reviews of current "traditional" telemedicine programs as well as "extreme" telemedicine applications are also presented. Current telemedicine programs, their development, licensure, accreditation, reimbursement policies, privacy policies, medicolegal aspects, and why many programs fail are also examined.

Patient data management and decision support systems
Description: Patient data management systems (PDMS), are innovative computer systems, which attempt to integrate administrative functions and clinical

decision making. Introducing this type of innovation tends to have far broader ramifications across the overall business domain. This course objective is to illustrate the resulting complexity of the relationship between this type of technology and organizational change through the investigation of as many facets as possible of the implementation of a PDMS in an intensive care unit (ICU).

Biomedical databases and biomedical research methodology

Description: This course will illustrate students extensively about relational databases as an alternative to textual databases. Topics include design and implementation of custom databases, modification or pre-existing databases, and database management.

Students will become well versed in SQL, including database construction, modification, and query design. Use of relational database as an analytical tool will be emphasized. This course will also provide an overview of the designing, conducting, analyzing, and interpreting phases of biomedical research procedure.

Teaching methodology

Distance Learning Modules: 8 modules of 16 weeks duration (8 × 10 ECTS each) and a final dissertation (1 × 20 ECTS), 100 ECTS in total. Each module includes weekly live lectures (lectures are recorded for asynchronous access), group consultation video conference meetings (tutoring students), module specific essays (collaborative/individual work)/virtual labs exercises (e.g., image processing, patient medical record, health informatics standards and project design, implementation and management) and module final examination.

Assessment methodology

The student evaluation performance will be based on individual work (exercises) (10%), teamwork (group exercises) (15%), midterm examination (25%), and final examination (50%).

Continuous professional education—Tele-Prometheus paradigm

Improvements in eHealth and technologies are creating the groundwork for a revolution in education, allowing learning to be personalized, collaborative, and learner centered learning (transforming the role of the teacher from disseminator to facilitator), following the principles of open education.

Emphasizing results driven approach Tele-Prometheus adopts the concept of triangulation of self-assessment, peer review, and quality of care indicators in shaping education and training priorities. In this changing paradigm, educators no longer serve as unique distributors of content, but they rather become facilitators of learning facilitators and tutors. Tele-Prometheus enables educators to do so by providing them with a set of online resources to facilitate the learning process.[13]

Objectives

The primary objective is the implementation and continuous upgrade of a novel tele-education platform, targeting health professionals, by means of enrichment of working environment with educational procedures.

Secondary objectives: (1) Transcript clinical needs into education processes; (2) integrate those processes into the clinical routine of healthcare professionals including clinical information systems; and (3) extend eLearning services to patients' relatives creating an integrated experience from hospital environment to the community.

Platform description

Training in demanding professional environments, such as that of an ICU, combines theoretical knowledge, practical skills, integration of information technology (IT) into clinical practice, and attitude formation tailored to extreme conditions. It is also a great challenge for the disciplines of Health, Education, and IT and Communication, given the high standards (zero tolerance for errors), high cost, shortage of time, and inefficiency of conventional educational methodology. Moreover, training of citizens for initial management of acute and life-threatening conditions (e.g., cardiac arrest), as well as the follow-up of chronically ill patients who live in the community after their ICU discharge are of great importance.

Services

The goal of Tele-Prometheus platform is the distance learning of health professionals, of patients and their families. To achieve this, it has been composed a technological platform using the latest technology tools and training systems for distance learning education of adults.

The infrastructure currently provides the following services:
- Website for health professionals, patients, and relatives

- Learning management system
- Management and search educational and information material
- Teleconference system (virtual classroom)
- Fully equipped rooms with teleconferencing systems (interactive whiteboard, teleconference system, audio system)
- Info-kiosks (interactive computers with easy navigation to sources of information)
- Time programmed system for educational purposes
- Audio and video system real time system (live streaming)

Short courses

Educational seminars are offered on the platform as short courses. The meaning of short has more to do with the work of the trainee rather than the duration of the training. The short courses are exclusively done by the instructors and consist of:

- Modern educational material (with certain principles and determined by qualitative criteria) that shared or posted to trainees on the platform. Trainees should study before the lecture.
- The lecture has a character of discussion and takes place from distance. The trainees can participate in groups (organized in rooms) or individually (from their PCs) at teleconference.
- Each teleconference may be accompanied by evaluations or questionnaires for instructors and trainees.
- The picture and the sound of the teleconference room can transfer everywhere so trainees can participate in virtual workshops.
- The final evaluation is done at specific examinations centers.

Educational seminars

Trainees can participate in specially equipped classrooms in which educational seminars can take place with high level education using interactive whiteboard. Professionals and experts from aboard can participate at seminars. Through our systems, presentations and image of medical device can be transferred.

Information points/kiosks

The trainees receive training injections in form of text, image, and video in their workplace through specially designed computers. Each information point can be time programmed with separate viewer program.

Info-kiosk for healthcare professionals

Modern computers with touch screens placed at information points in the ICU offering a window to knowledge for professionals whose decisions and actions have zero tolerance to errors.

The training scenarios of info-kiosk service divided into:

- Health professionals use the touch screen to navigate and retrieve simply and fast guidelines, protocols, educational materials, etc.
- The directors of departments plan their educational injections in the form of text, images, and videos at work of their partners. Each kiosk can follow separate viewer.

Info-kiosk for patients and their relatives

One of the main objectives of the project Tele-Prometheus are education/information of patients and families about (1) the structure and the activities of the ICU; (2) the most common diseases of patients which are hospitalized in the ICU; (3) medical terms; and (4) medical devices.

Direct information of relatives through the platform extends bridges across the communication gap with health professionals and saves considerable time of visiting relatives spent by health professionals for general information.

The personalized education of patients and their families intend to improve the post-hospital period and particularly during the critical phase of rehabilitation and reintegration into the community.

Tele-consultation

Professionals can communicate with experts from all over the world and invite them to a virtual room for advice or evaluation (biosignal transmission, diagnostic images, and video).

Current and future eLearning technologies in healthcare education

Implementing an eLearning environment to successfully deliver the curriculums described in the previous paragraphs for both students and professionals is non-trivial and introduces several challenges. No single method, technique or methodology can meet all these challenges but rather a graceful synthesis of current and future technologies/techniques can be used. A complete learning environment must provide the tools for trainees to acquire competencies in all learning counterparts (knowledge, skills, and attitudes).

Knowledge

Escaping the traditional classroom teaching, ICTs have played an important role in education introducing advanced online learning environments, Learning Management Systems (LMS), such as Moodle and Moodle based tools.[14,15] These systems provide many tools to deliver the knowledge counterpart of a curriculum. LMS provide integrated support for six different activities: creation, organization, delivery, communication, collaboration, and assessment. Features such as interaction, feedback, conversation, and networking are some of the available functionalities used by learning platforms. Furthermore, LMS provide a variety of learning activities such as creation, organization, delivery, communication, collaboration, and assessment (and self-assessment). These systems also provide a set of configurable tools to enable the creation of online courses, working groups, and learning communities. In addition to the pedagogical aspect, these systems have a set of features for evaluation and self-evaluation activities for students and teachers. These features can be used to assess the knowledge competencies acquired from an online course and help the tutor identify his/her students' learning needs.

Skills

Although the knowledge counterpart can be achieved within the context of an online course, delivering the skills counterpart requires the introduction of psychomotor learning (a combination of cognitive functions and physical movement). Except a minor set of learning topics in Education and especially healthcare education, the skills counterpart requires face to face education. Introducing face to face (not necessarily real presence) meetings/workshops/lectures into the eLearning curriculum forms a hybrid curriculum often referred as **Blended Training**. The tutor utilizes the full functionality of the LMS to prepare the trainees for the face to face event where they will get to participate in activities to acquire or improve their skills on specific topics.

Blended training enhances learner experience and the expected intended learning objectives in the following aspects. For example:

1. Using self-assessments and assessments the tutor can identify the real educational needs of his/her trainees and prepare either online or face to face meetings to address these needs.
2. Online meetings can be used for open discussions or tele-consultation either to individuals or groups for specific topics of the curriculum.

3. Online lectures either in form of live or recorded webinars can be used to better communicate curriculum parts.

4. Online workshops and software applications especially those with collaboration features can be used to acquire or improve skills in a variety of topics such as project and time management, organization and management, healthcare analytics, medical device operation, etc.

Attitudes

The attitude counterpart often is the most difficult to achieve. Generally, just because we know something, or we have the skills to accomplish an activity do not mean that we do it or we do it well. What we do often is not aligned with our knowledge or skills. To address this challenge, tutors teach by example. This is most effective within the working environment and requires that the tutor is well prepared to perform everyday work aligned to the curriculum. The tutor also initiates ad-hoc discussions with the learners to point out what they are doing wrong or to suggest better ways of doing things. This counterpart of the curriculum can be further strengthened utilizing learning technologies such as time scheduled videos within the working environment to stimulate positive thinking, gamification within the learning curriculum to further involve the learning into more learning activities, and interactive case scenarios to demonstrate the reasoning behind specific attitudes.

References

1. Topol EJ. The patient will see you now: the future of medicine is in your hands. *J Clin Sleep Med* 2015;**11**(6):689–90.

2. Schiza EC, Neokleous KC, Petkov N, Schizas CN. A patient centered electronic health: eHealth system development. *Technol Health Care* 2015;**23**(4):509–22.

3. *Tele-Prometheus project, Intensive Care Unit–Nicosia General Hospital, Project Deliverable 3.3.1 [Internet].* Available from: https://www.intensivecare.com.cy/teleprometheus/images/Documents/ParadoteaErgou/methodologia-ekpedeusis.pdf.

4. Vaitsis C, Spachos D, Karolyi M, Woodham L, Zary N, Bamidis P, et al. Standardization in medical education: review, collection and selection of standards to address. *MEFANET J* 2017;**5**(1):28–39.

5. *University of Cyprus, School of Medicine, eHealth and Medical Informatics Course–CS041 [Internet].* Available from: https://moodle.cs.ucy.ac.cy/course/view.php?id=94.

6. *AFMC (The association of faculties of medicine of Canada), e-Learning initiatives [Internet].* Available from: https://afmc.ca/medical-education/e-learning-initiatives.

7. Ludwick DA, Doucette J. Adopting electronic medical records in primary care: lessons learned from health information systems implementation experience in seven countries. *Int J Med Informat* 2009;**78**(1):22–31.

8. *Guidelines on minimum/non-exhaustive patient summary dataset for electronic exchange in accordance with the cross-border directive 2011/24/EU, Release 1 [Internet].* Available from: https://ec.europa.eu/health//sites/health/files/ehealth/docs/guidelines_patient_summary_en.pdf.

9. *Integrating the Healthcare Enterprise (IHE), IHE IT Infrastructure White Paper, Health IT Standards for Health Information Management Practices [Internet].* Available from: http://ihe.net/uploadedFiles/Documents/ITI/IHE_ITI_WP_HITStdsforHIMPratices_Rev1.1_2015-09-18.pdf; 2015.

10. Neofytou MS, Neokleous K, Aristodemou A, Constantinou I, Antoniou Z, Schiza EC, Pattichis CS, Schizas CN.;1; Electronic health record application support service enablers. In: *Engineering in medicine and biology society (EMBC), 2015 37th annual international conference of the IEEE*; 2015,p. 1401–1404.

11. Schizas CN, Healthcare ecosystem: the patient will see you now [Internet]. *eHealth Forum 2017, Digital Health Conference.* Available from: https://www.youtube.com/watch?v=m7uW1Ue0bEg; Athens, 2017 [cited 11 December 2017].

12. *Open University of Cyprus, faculty of pure and applied sciences, applied health informatics module [Internet].* Available from: http://www.ouc.ac.cy/web/guest/s2/progrspoudon/hin/desc.

13. Ruiz JG, Mintzer MJ, Leipzig RM. The impact of e-learning in medical education. *Acad Med* 2006;**81**(3):207–12.

14. Haftor D, editor. Information and Communication Technologies, Society and Human Beings: Theory and Framework (Festschrift in honor of Gunilla Bradley): Theory and Framework (Festschrift in honor of Gunilla Bradley). IGI Global; 2010 Jul 31.

15. Antoniades A, Nicolaidou I, Spachos D, Mylläri J, Giordano D, Dafli E, et al. Medical content searching, retrieving, and sharing over the internet: lessons learned from the mEducator through a scenario-based evaluation. *J Med Internet Res* 2015;**17**(10).

Index

Note: Page numbers followed by "f" indicate figures, and "t" indicate tables.

Printed in the United States
By Bookmasters